CONTENTS

NOTES ON CONTRIBUTORS

Stuart Aitken is Professor of Geography at San Diego State University, where he is also the director of the Center for Interdisciplinary Studies of Youth and Space; and an Adjunct Professor in the Department of Geography at the National Technical University of Norway. His research interests include film and media; children, families, and youth; critical visual approaches in GIS; and qualitative and poststructural methods in geography. Stuart has published widely in academic journals in urban and social geography and GIS, and is the author or editor of multiple scholarly texts contributing to social geography and research methodologies in geography. He is currently the North American editor of *Children's Geographies*.

Meghan Cope is an Associate Professor in the Geography Department at the University of Vermont in Burlington, Vermont, USA. Her research interests are in the local experiences and constructions of place among marginalized social groups of North American and European cities. Most recently, Meghan has been working on research with young people through the Children's Urban Geographies project. She has also published widely on qualitative research methods and combines this with an interest in critical GIS to engage in new conceptual and practical developments, particularly through work with students.

Jon Corbett is an Assistant Professor in the Community, Culture and Global Studies Unit at UBC Okanagan, British Columbia, Canada. Jon's community-based research investigates cartographic processes and tools that are used by communities to help express their relationship to, and knowledge of, their territories and resources. Jon has worked with indigenous communities in Australia, Indonesia, and the Philippines, and since 2004 with several First Nations communities in British Columbia. Recently Jon has been working with the United Nations International Fund for Agricultural Development in Albania, Kenya, Mali, and Sudan.

Jim Craine is an Assistant Professor in the Department of Geography at California State University–Northridge, California, USA. Recent research includes work on affective geovisualization and emotional cartographies. He is also a co-editor of *Aether: The Journal of Media Geography*.

Sarah Elwood is an Associate Professor in the Department of Geography at the University of Washington, in Seattle, Washington, USA. Her research interests intersect critical GIS, urban political geography, qualitative methods, and participatory action research. In teaching, her contributions include developing curricula for incorporating participatory and community-engaged research into undergraduate GIS education. Most recently, Sarah has been working on a collaborative research, teaching, and community outreach project that focuses on the use and impacts of geographic information systems and GIS-based spatial knowledge in neighborhood revitalization and in urban planning and problem solving.

Jin-Kyu Jung is an Assistant Professor in the Department of Geography at the University of North Dakota, USA. His main academic interests are in GIS and qualitative research, community studies, critical urban geography and planning, and mixed methods research. He has been conducting research on how to more fully integrate qualitative and quantitative approaches in mixed methods research with GIS, most recently through two case studies involving children's and adults' perceptions of their neighborhoods in Buffalo, NY. Exploring further applications of qualitative GIS as an analytical framework to study diverse urban issues is his long-term research goal.

LaDona Knigge is an Assistant Professor in the Department of Geography and Planning at California State University–Chico, California, USA. Her interests include community gardens, public space, qualitative GIS, and local food systems, with particular emphasis on fostering and promoting civic agriculture and relocalizing food systems in ways that build both community and economic and ecological sustainability.

Marianna Pavlovskaya is an Associate Professor in the Department of Geography at Hunter College and in CUNY Graduate Center, in New York, USA. Her research focuses on the constitution of class and gender in postsocialist Moscow and New York City, rethinking neoliberal transition in Russia, and critically rereading geospatial technologies. She uses GIS in qualitative and quantitative research projects and is interested in its effects as a particular and very powerful representational tool.

Giacomo Rambaldi is senior programme coordinator at the Technical Center for Agricultural and Rural Cooperation (CTA) in The Netherlands. He has 26 years of professional experience in developing countries where he worked for a number of international organizations. Giacomo has been promoting good practice in the domain of community mapping via various channels including his websites www.iapad.org and www.ppgis.net. Areas of professional interest include visualizing indigenous spatial knowledge for improving communication, facilitating peer-to-peer dialog, and managing territorial conflicts; collaborative natural resource management; participatory spatial planning; networking; and web publishing.

Nadine Schuurman is an Associate Professor of Geography at Simon Fraser University in British Columbia, Canada. Her research is at the intersection of human geography and geographic information science. She has parallel research interests and expertise in theoretical issues in GIS as well as GIS health informatics. She has conducted research on population health, spatial epidemiology, hidden homelessness, groundwater data, data integration and standardization, ontologies, and metadata.

Matthew W. Wilson is a PhD Candidate in Geography at the University of Washington in Seattle, Washington, USA. His interests include critical GIS, science studies, and subject formation, and he is conducting research examining the cultural and political work of community mapping projects completed by a Seattle-based non-profit. He is interested in how geographic information technologies enable these kinds of neighborhood assessment endeavors, and how these geocoding activities mobilize particular notions of 'quality of life' and 'sustainability'.

LIST OF FIGURES AND TABLES

ACKNOWLEDGEMENTS

From its earliest imagined existence to the finished manuscript, this book has emerged amidst two job changes, two promotion and tenure processes, five household moves, periods of intensive elder, partner, and child care, and countless telephone conversations. So naturally, there are many people to thank. First, we thank the authors for their outstanding contributions, their creative engagements with the notion of qualitative GIS, and their hard work in bringing the final manuscript together. As well, we are grateful to the 19 anonymous reviewers for their thoughtful feedback on the initial chapter submissions. Robert Rojek at SAGE offered us persistent encouragement to propose this text and unflagging support and advice through the process of bringing it to fruition. Sarah-Jayne Boyd, also at SAGE, has been a tremendous source of help in navigating the details toward publication. Milissa Orzolek at the University of Washington took on the Herculean task of copyediting and formatting the final manuscripts, under tremendous time pressure. We are grateful to all of them.

We both wish to acknowledge the invaluable role that the US National Science Foundation's CAREER award program played in positioning us for this project. These awards support junior faculty to undertake a five-year program of interlinked research and education activities, a structure that encourages emerging scholars to think creatively about productive substantive, methodological, and pedagogical collisions *and* provides the sustained support needed to realize these innovations. For these reasons, while *Qualitative GIS* was not conceived as part of our respective CAREER projects, it is still a direct result of the ideas, imaginations, and activities inspired by them. We especially thank Tom Baerwald for his support throughout the various challenges and opportunities of our respective CAREER projects (Cope, BCS-9984876, 2001–2006; Elwood, BCS-0652141, 2003–2008).

Finally, we thank treasured friends, colleagues, and family whose ideas, commitments, patience, and care have helped us bring this book into being: Alan Coté, Lisa Faustino, Geneva and Celia Coté, Jim and Susan Cope, Dale Ferguson and Eric Hudson, Marj and Norm Elwood, Helga Leitner, Deb Martin, Vicky Lawson, Rachel Pain, Sara Kindon, Jennifer Pierce, Nadine Schuurman, Eric Sheppard, Chris Brehme, Glen Elder, David Mark, Jin-Kyu Jung, LaDona Knigge, Maureen Sioh, and Lynn Staeheli.

Meghan Cope
Sarah Elwood

1

INTRODUCTION: QUALITATIVE GIS: FORGING MIXED METHODS THROUGH REPRESENTATIONS, ANALYTICAL INNOVATIONS, AND CONCEPTUAL ENGAGEMENTS

Sarah Elwood and Meghan Cope

The title of this volume may, for some readers, suggest contradiction, incongruity, or juxtaposition. From its inception, this project has been met with some measure of all three, and with several persistent questions. Why qualitative GIS? Why a mixed methods approach? Why not just 'mixed methods GIS'? What is 'qualitative' in the context of a digital technology? In spite of such questions, efforts to integrate qualitative data and techniques with GIS have been building in recent years (Kwan and Knigge, 2006). Multimedia GIS approaches embed sketches, mental maps, audio, video, or photographs into GIS, often to represent non-cartographic forms of spatial knowledge, such as emotion (Al Kodmany, 2002; Kwan, 2007; Shiffer, 2002; Weiner and Harris, 2003). A growing number of researchers use GIS-based spatial analysis in concert with methodologies more familiar to qualitative researchers, such as focus groups, ethnography, interviewing, or participatory action, thus strengthening research findings by bringing together these different ways of knowing (Cieri, 2003; Dennis, 2006; Pain et al., 2006; Weiner and Harris, 2003). Others are developing ways to use GIS as part of a suite of analysis techniques drawn from qualitative research, whether by adapting GIS software or by using it to carry out inductive interpretive visualization (Knigge and Cope, 2006; Kwan and Lee, 2004; Matthews et al., 2005). Building upon these and other examples, this collection is intended to frame the emerging field of qualitative GIS, profiling the range of ways in which researchers and practitioners are integrating GIS with qualitative research.

Qualitative GIS is one of several approaches to geographic information systems that emerged in response to critiques in the mid 1990s that cast GIS as rooted in positivist epistemologies and most suited for quantitative techniques associated with spatial science (Lake, 1993; Pickles, 1995). These critiques also raised concerns about the difficulty of incorporating non-cartographic spatial knowledge into conventional GIS, and the ensuing potential for exclusion and disempowerment (Harris and Weiner, 1998; Sheppard, 1995). Responding to these critiques, many researchers have taken on GIS in new ways, working to incorporate multiple data and forms of knowledge, extend its representational capabilities to incorporate non-cartographic information, support quantitative and qualitative forms of analysis, and illustrate that multiple epistemologies may be part of GIS-based research. These approaches, including public participation and participatory GIS, feminist GIS, and critical GIS, share an understanding of GIS as more than only quantitative in the forms of data that may be included, the analyses that may be supported, and the representational practices that may be fostered. These propositions have paved the way for qualitative GIS.[1]

At the level of the technology, it is increasingly simple in most desktop GIS software to store multimedia forms of spatial knowledge (photographs, sketch maps, narrative descriptions) in a GIS database or to hyperlink to them in other locations. But qualitative GIS is more than this. It stands apart from conventional GIS practices, as well as some of the approaches described above, because of the way it engages and conceives of GIS. Specifically, qualitative GIS assumes that geographic phenomena, their relationships, and their meanings are produced and negotiated at many different moments in GIS development and application: in spatial data, in data structures, in spatial analysis techniques, in the meanings fostered or foreclosed in GIS-based maps and applications. As this collection will illustrate, this assumption enables approaches to GIS that foster new collisions with qualitative analysis techniques, qualitative forms of data, and new conceptualizations of how meaning is negotiated in and through different aspects of GIS, including its software, data structures, and visualizations (cartographic and otherwise). Further, qualitative GIS is predicated upon multiple representations and modes of analysis, hybrid epistemologies, and researchers' critically reflexive efforts to draw on these multiplicities in ways that enable more robust explanation. In the remainder of this introductory chapter, we further detail this framing of qualitative GIS, outline some of its foundations and practices, and situate its origins in and contributions to geographers' long-standing reliance on mixed methods.

WHAT IS GIS AND WHAT CONSTITUTES 'QUALITATIVE'?

Identifying the emerging field of qualitative GIS, characterizing its differences from other approaches to GIS, and explaining its intellectual and political significance require some initial discussion of what is meant by 'GIS' and what is meant by 'qualitative'. GIS research and practice over the past decade constitutes GIS in multiple ways, and geographers' conceptions of qualitative research are similarly multi-faceted. This complexity produces many ways of bringing GIS and qualitative research together, and generates a diversity of sites and practices through which qualitative GIS might emerge.

For more than a decade, GIS research and critiques have conceptualized GIS in several ways (Pickles, 1995). Geographic information systems are, in one understanding, digital technologies for storing, managing, analyzing, and representing geographic information. Typically, such a system consists of data models; structures for representing geographic entities and their characteristics in digital form; data structures for storing these data; the data themselves (together with the ontologies, categorization schemes, and other elements that are part of these representations); software for query, retrieval, analysis, and mapping; and the hardware used to support these functions (Chrisman, 2002).[2] But simultaneously, GIS is understood as a collection of practices for producing and negotiating geographic knowledge through the representation and analysis of spatial data. These practices are constituted by GIS software producers and the predominantly private sector industry that creates GIS software; geography and other academic disciplines that create and validate certain ways of encoding spatial data in a GIS, or using GIS in research; and the ever-diversifying 'community' of GIS users who create various GIS practices (Pickles, 1995). From this perspective, GIS is constituted through its representational, analytical, and epistemological approaches, all of which are understood to be shaped by the social, political, and disciplinary norms and institutional practices from which they emerge. This conceptualization of GIS owes much to the efforts of researchers to respond to claims about an 'inherent' positivism in GIS and explain how the social and political impacts of GIS might be produced (Pavlovskaya, 2006; Sheppard, 1995), as well as to feminist geographers' critical reflections upon the social, political, and institutional construction of knowledge in research (Lawson, 1995; Mattingly and Falconer Al-Hindi, 1995; Moss, 1995). Our notion of qualitative GIS is rooted in this hybrid understanding of GIS as technology, methodology, and situated social practice.

What, then, counts as 'qualitative' in our account of qualitative GIS? First, data or forms of evidence in research may be qualitative. Qualitative data are not simply those data that are non-numerical. Rather, we argue that data may be qualitative in part by virtue of the rich contextual detail they provide about social and material situations. Ethnographic interviews, for instance, tend to elicit responses from interviewees that describe conditions, relationships, and processes in detail. If we were to ask neighborhood residents how a local community development initiative has changed their neighborhood, some might describe changes to the built infrastructure, some might relate stories of how local residents' ability to impact neighborhood planning has changed, and others might describe how new residents have moved in, altering social relations in the neighborhood. The interviewees' responses are qualitative data because each narrative likely communicates rich descriptive detail about these shifting social and material conditions and processes.

But it is not only the presence of rich contextual descriptive detail that constitutes data as qualitative. Rather, data may also be qualitative if they contain or provide interpretations of the situations or processes that they describe. For example, in the ethnographic interview responses described above, the words chosen and the stories related by the interviewees will almost certainly evidence their own interpretations or the meanings they draw from particular events or conditions. One person may describe a newly constructed grocery store in the neighborhood as an example of positive changes that improve food security for low-income households in the community, while another may describe the same site and characterize it as a harbinger of gentrification – a potential threat to these same low-income households.

These forms of evidence are qualitative because they not only encapsulate a description of material change, but also offer interpretations of the meanings or impacts of that change.[3] As well, these data are qualitative because we can use them to understand situated and negotiated knowledge. That is, the different meanings developed in the two interviewees' characterizations of the grocery store are surely *also* being contested and negotiated among neighborhood residents, and so these two pieces of evidence provide a researcher insights into the social and political situations of the neighborhood. Further, the researcher might examine the identities, experiences, and interests of the two interviewees to understand how these factors affect the meanings they provide – understanding how their respective explanations provide differently situated knowledges.

Thus, it is not only data that may be qualitative, but also analysis. That is, we also understand as qualitative those forms of analysis that are intended to draw out the situated interpretive detail of qualitative data. Analytical techniques such as grounded theory, discourse analysis, or content analysis, for example, work with qualitative forms of evidence to tease out their negotiated meanings and situated knowledges. Techniques such as coding (systematically categorizing data to identify themes and patterns) and the triangulation of multiple data sources are associated with qualitative research because they enable researchers to examine the contradictions, commonalities, and nuances of data that are rich in contextual and process-based detail. Practices such as the iterative or recursive examination of multiple forms of evidence in conversation with one another are also characteristic of qualitative analysis. And finally, as has been well developed in feminist researchers' writing on qualitative methods, a hallmark of qualitative analysis is its critical reflexivity upon the knowledge production process, specifically how research designs, forms of data, analysis techniques, disciplinary politics around epistemologies, and research relationships (such as the position of the researcher *vis-à-vis* the participants) tend to produce particular forms of knowledge, conclusions, politics, and power relations (Mattingly and Falconer Al-Hindi, 1995; Moss, 2003; Nast, 1994).

These understandings of GIS and qualitative forms of data and analysis foreshadow the multiplicity of engagements that might constitute qualitative GIS. We understand as qualitative GIS those approaches that seek to integrate qualitative forms of data into GIS, develop and support qualitative approaches to building knowledge and explanation with GIS, use GIS in research that emerges from multiple or hybrid epistemologies, and theorize previously unrecognized forms of social knowledge that may be present in GIS applications. These approaches are quite different in the way they bring together GIS and qualitative research, but all go beyond treating qualitative methods as 'add-ons' to essentially quantitative projects rendered in a GIS. Instead, they offer substantive shifts toward framing questions, collecting data, analyzing results, and representing findings in a truly integrated way. They intersect GIS and qualitative research with the goal of integrating multiple forms of evidence or ways of knowing, in order to explain *how* spatial knowledge, patterns, relationships, and interactions are produced, and with what sorts of social and political impacts.

POSITIONING QUALITATIVE GIS AS A MIXED METHODS APPROACH

The core commitment of qualitative GIS to *integrating* multiple forms of knowledge and the findings from various techniques is also at the heart of mixed methods research. This integrative way of building robust explanations in research is what positions qualitative GIS as a mixed methods approach. Distinguished from 'multiple methods' projects in which different methods are practiced in parallel, *mixed* methods projects weave together diverse research techniques to fill gaps, add context, envision multiple truths, play different sources of data off each other, and provide a sense of both the general and the particular. In these approaches, insights gained from one technique, subject group, or data source may be examined recursively with other findings, and the path of research may be shifted in response (Cresswell, 2003; Jiang, 2003; Robbins, 2003). Different techniques may produce complementary explanations for phenomena, while other times (and equally valuably) they may produce contradictory explanations, leaving the task of understanding how and why these multiple versions of 'truth' intersect (England, 1993; Nightingale, 2003; Pavlovskaya, 2002; Rocheleau, 1995; Tashakkori and Teddlie, 2003). Mixed methods have been especially important in geography research, given the strong presence of research questions that require investigating interrelated human and physical processes, understanding cognitive and social processes, or examining interscalar relationships and processes. For example, accounting for the role of structures such as political economy or gender, while also understanding how and why they may play out differently in various contexts, is precisely the sort of integrative project that would necessitate relying on multiple forms of evidence and diverse epistemologies (ways of knowing).

Mixed methods approaches are rooted in several assumptions about knowledge and epistemologies in research that we suggest are also critical for qualitative GIS. First, mixed methods research tends to treat knowledge as always partial (no one can know the 'whole truth') and situated (knowledge depends on our situations and positions), whether it is the forms of evidence that researchers 'gather' or the knowledge that they produce in their analysis, interpretation, and representation of these data. From this perspective, differently situated knowledges and multiple ways of examining evidence can inform more robust understandings of complex processes or phenomena. Second, mixed methods research is premised on the notion that epistemology and methodology are related, but that this relationship is neither fixed nor singular. A realist, positivist, or constructivist epistemology need not prescribe a given methodological orientation, or only one approach (qualitative or quantitative). Indeed, some scholars contend that mixed methods research forwards unique hybrid epistemologies. Inherent in efforts to bring together multiple ways of knowing, they contend, is an assumption that multiple epistemologies may be valid ways of fostering understanding and explanation *for particular purposes and in specific circumstances* (Elwood, 2009a; Knigge and Cope, 2006; Maxcy, 2003). Finally, a great deal of mixed methods research rests on the assumption that the knowledge making we do in research is inherently political. The manner in which researchers interpret tensions or contradictions among data or methods and weave together different approaches has social and political consequences, especially because different forms of data, representation, and analysis are frequently afforded different levels of intellectual and political authority (Elwood, 2006; McCann, 2008; McLafferty, 2002). All of these propositions underscore the importance of researchers' critical agency in bringing together multiple epistemologies, modes of analysis, and forms of knowledge.

Emerging efforts to intersect GIS and qualitative research share many of these same commitments. The chapters in this collection illustrate the persistent methodological and epistemological multiplicity of qualitative GIS, and the authors' efforts to work with GIS in ways that foster debate, multiple readings, and iterative interpretation, with sustained critical reflection upon methods and outcomes. Qualitative GIS involves conceiving of GIS as social and political practice, as discipline- and industry-inscribed ways of making knowledge, and as an assemblage of hardware, software, data structures, and procedures for working with digital spatial data. This conceptualization of GIS allows us to consider GIS-based knowledge production as something that is occurring explicitly and implicitly in several sites: spatial data, data structure, spatial analysis, cartographic representation, and the application or use of any of these phenomena in social and political practice. This multi-layered reading of how knowledge is produced in GIS suggests that the 'qualitativeness' of qualitative GIS may be advanced in many ways: by integrating qualitative and quantitative representations of spatial knowledge; by engaging multiple modes of analysis; and by incorporating GIS and digital spatial data in research that is premised upon multiple epistemologies.

Building upon these notions of multiple or hybrid epistemologies, qualitative GIS emphasizes the infinitely creative and political possibilities of bringing together multiple ways of knowing and making knowledge. Qualitative GIS is in part defined by purposeful and critical engagements with different aspects of *fixity* that may be part of GIS practice. Because of its basis in computing and the centrality of cartographic representation and analysis, GIS tends to 'fix', to pin down, knowledge, representations, or meanings at particular moments. A map, for instance, may be fixed at the moment of hard copy production, or fixed at the moment that it is presented with a specific interpretation, or fixed at the moment that it is witnessed or interpreted by a map user. Geographic knowledge is fixed when it is encoded into a GIS-based data structure as digital spatial data. Any spatial knowledge, relationship, or analysis must be able to exist in 'code space' – captured, represented, or expressed in programmable digital forms – if it is to be part of a GIS (Schuurman, 2006). But the varied practices that constitute qualitative GIS share an assumption that while some kinds of fixity are inherent and unavoidable in GIS, there exists a great deal of room for strategic deployments of this fixity, and for iterative adaptations of fixed representations or practices. Qualitative GIS critically examines the implications of these moments of fixity, and intentionally engages them for specific purposes. Imperative in this view is the understanding that any effort to fix meanings (or to disrupt them) is inherently power-laden, inseparable from the performative, representational, or analytical practices through which these meanings are produced.

From these conceptualizations of GIS and its knowledge-making possibilities, there are many ways that GIS-based knowledge production might *be* qualitative, and many positions in the layerings of GIS from which a qualitative GIS might be activated. Amidst the diversity of emerging practices, we discern three closely related approaches to qualitative GIS. Some of the interventions and innovations that mesh GIS and qualitative research focus upon GIS-based representations, including spatial data, maps, or other visual representations. Others focus upon the forms of analysis that can be carried out in connection with GIS, considering how it might be part of knowledge production practices more common to qualitative research. Still other forms of qualitative GIS are reflexively conceptual or theoretical, examining GIS through theoretical frameworks that can highlight or provide insight into qualitative forms of knowledge or knowledge making that are part of GIS research and practice, yet often overlooked. This tripartite framing of qualitative GIS approaches specifies how this emerging field goes about developing different aspects of 'the qualitative' in research with and about GIS.

QUALITATIVE GIS: REPRESENTATIONS, ANALYTICAL INTERVENTIONS, AND CONCEPTUAL ENGAGEMENTS

As noted earlier, qualitative GIS owes much to more than a decade of creative and critically reflexive engagement with GIS on the part of diverse scholars, especially efforts to challenge assumptions about the association of GIS with quantitative methods or a positivist epistemology (Kwan, 2002; Pavlovskaya, 2006; Schuurman, 2000; Sheppard, 1995). These efforts to decouple assumed linkages between GIS and specific methodologies or epistemologies have created space for qualitative GIS because they began to rewrite the discipline-inscribed narratives that create and reinforce such associations. Thus, we open with a chapter in which Pavlovskaya offers such a rewriting. She positions GIS amidst epistemological struggles in geography, to show how and why its basis in cartography and computing became linked to spatial science and quantitative research, and traces the critiques that have enabled its reconstruction toward a wider range of methodologies and epistemologies. In challenging the assumed quantitative orientation of GIS, she shows that GIS in fact comprises complicated layerings of analytical, representational, and political practices that are more than only quantitative. Her discussion of openings for qualitative representations or ways of knowing foregrounds the approaches to qualitative GIS around which the following three sections of the text are structured: 'Representations', 'Analytical Interventions and Innovations', and 'Conceptual Engagements'.

The first section profiles qualitative GIS approaches that are advanced through representations, whether in spatial data, maps, or other visual representations that can be produced with or embedded into a GIS. Here, we understand representations in the context of a GIS to be those artifacts that stand in for some 'real world' geographic phenomena or relationships – a spatial dataset, a map, a single data attribute characterizing some geographic phenomenon – but also the multiple meanings negotiated through these artifacts. These approaches to a qualitative GIS involve stretching the existing limits of GIS-linked representations to incorporate qualitative information, as well as their untapped potential to incorporate and produce multiple forms of knowledge. A qualitative GIS is attentive to how these representations can produce openings and closures for researchers to include multiple politics, experiences of the world, or ways of engaging the same information.

Schuurman's chapter is premised on the notion that the spatial data in a GIS are themselves representations of characteristics in a complex 'real world'. She illustrates a persistent problem that emerges when these characteristics are represented as data, through measurements, categorization schemes, and attributes. In this representational moment of translation, important qualitative information is lost. The data no longer bear information about the original purpose of their collection, the institutional imperatives that may have influenced the categories chosen, or the nuanced semantics through which 'real world' characteristics are encapsulated in their categorization schemes. These contextual details are imperative for understanding how spatial data may or should be analyzed, integrated with other data, or applied to inform research questions or policy decisions. Further, she develops an approach for including some of this valuable qualitative knowledge with spatial data, by expanding existing metadata structures that are used in spatial databases to incorporate information about the data.

Elwood focuses on a different representational aspect of GIS: maps as representations of complex spatial knowledge. She suggests that these representations are flexible and fluid, holding the potential for map makers and map users to interpret and reinterpret them to produce different meanings. As such, she suggests that these cartographic representations emerging from GIS practice are also constitutive. As maps are presented and performed, they shape the meanings, identities, and characteristics that individuals and groups may assign to individual places, and even produce the places that are there to know.

In Chapter 5, Corbett and Rambaldi also focus on the capabilities of GIS-based representations to encompass and foster multiple meanings, experiences, and knowledges. Drawing examples from community mapping initiatives in the global South, they show that the capacity of GIS to include diverse local knowledge and experiences may be expanded by weaving together GIS-based maps with an array of other visual media, including sketch maps, three-dimensional models, or paintings. They emphasize that the emergence of complex social knowledge from community mapping depends not just on these multimedia representations, but upon richly interactive processes of knowledge production in community mapping – such that participants examine, negotiate, and communicate their spatial knowledge and experiences together as they are producing maps, models, or other visual representations. Their account of community mapping contributes to a qualitative GIS repertoire through its critical attention to how multiple situated knowledges might be included and fostered through approaches to GIS that push beyond its conventional representational practices.

The second section focuses on qualitative GIS practices that are activated through the modes of analysis that are employed. These analytical interventions integrate GIS as part of the inductive interpretive techniques that are often used to tease out the situated contextual meanings of data, especially qualitative data. Qualitative GIS that is fostered through analysis tends to integrate multiple forms of data – maps, photographs, interview transcripts, field notes, sketches – in order to explore the differences, contradictions, and points of agreement among and between them. Here we see again a strong mixed methods influence, in the emphasis upon integrating multiple forms of evidence and multiple ways of knowledge to gain greater insights than might be possible through more singular perspectives or approaches.

One such analysis-rooted approach to qualitative GIS is Knigge and Cope's (2006) 'grounded visualization'. Grounded visualization is an iterative reflexive engagement with multiple forms of data (GIS-based spatial data and maps but also photographs, field notes, and other evidence) that integrates exploratory visual methods with analysis practices drawn from grounded theory. In their chapter for this collection, they show how a grounded visualization approach can support a scale-sensitive analysis that is attentive to the scale of *spatial data*, the scale of *social and political processes*, and the scale of *cartographic representations*, in ways that foster greater insight into the situations that are evidenced in these data, processes, and representations. Drawing examples from a study of urban community gardening, they illustrate how the iterative, recursive engagement with differently scaled data, processes, and representations that is part of a grounded visualization approach can illuminate contradictions and tensions over 'vacant' urban land and develop stronger explanations of the institutional, political, social, and economic structures and relationships that produce them.

Knigge and Cope's chapter shows us that GIS may be integrated with qualitative analysis without necessarily altering GIS software itself. However, the companion chapter in the second section illustrates that software-level interventions are another productive way of integrating qualitative analysis techniques with GIS. Jung presents ways of incorporating qualitative forms of data *and* qualitative data analysis techniques into GIS software and data structures. He describes several techniques he developed to link GIS software with computer-aided qualitative data analysis software (CAQ-DAS), in this case using the package ATLAS.ti. Jung's innovations enable the georeferencing of qualitative representations such as photographs or sketches so that they may be stored directly in GIS data structures, and storage of qualitative codes assigned by the researcher in the process of analyzing these data. These qualitative codes are used to connect the GIS software with qualitative data analysis software, such that the researcher has simultaneous access to the extensive data storage, representation, and analysis capabilities of both systems and can use these linkages to create analyses that are enmeshed at new levels.

The final section of the book outlines a third important strand of qualitative GIS: reflexive conceptual engagements with GIS research and application. These contributions do not center upon the practical considerations of integrating qualitative data or analysis techniques with GIS, or on the everyday challenges of data collection, analysis, synthesis, and representation. Rather, they contribute to qualitative GIS by developing new conceptual frameworks that illuminate previously unexamined aspects of the knowledge and knowledge making that are part of GIS research and practice. Some of these conceptual engagements directly examine qualitative GIS, while others illuminate qualitative aspects of more conventional GIS research and practice, using new conceptual readings of GIS to recover and uncover some of its tacitly qualitative elements.

Aitken and Craine's chapter offers the latter sort of conceptual engagement, illustrating how reading GIS practice through new theoretical frames can reveal ways of knowing that are ever-present in GIS but have not typically been recognized. Their *affective geovisualization* technique examines GIS-based visual representations with concepts drawn from non-representational theory, to show how these visualizations activate non-representational ways of knowing, such as emotion or affect. Their way of engaging both the representational and the non-representational in GIS is a promising response to recent calls for qualitative methods to do more than only read representations (such as maps, interview transcripts, or photographs) as texts that negotiate meanings (Crang, 2005). Affective geovisualization also contributes to qualitative GIS by highlighting the productive flexibility of GIS for fostering multiple ways of knowing.

Wilson's chapter offers a reflexive conceptual engagement with qualitative GIS itself. He situates qualitative GIS within a genealogy of different critical GIS approaches, reading these trajectories with concepts from feminist theory and method, especially feminist studies of science and technology. This reading of qualitative GIS research enables Wilson to characterize its unique approach to knowledge production and research practice – a 'techno-positionality' that is attentive to the influence of institutional and disciplinary practices, to the active agency of the researcher within the ethnographic relationships of GIS in practice, and to the power and potential of reworking GIS software and data structures. Techno-positionality is a critically reflexive interaction with GIS that engages 'the machine' on its own terms, while simultaneously seeking to create openings for new ways of knowing and for overlooked forms of knowledge making that might be present in GIS research and practice. Wilson urges the development of a qualitative GIS that continues to interrogate how engagements with GIS in research are enabled differently through multiple positions, with different implications for knowledge creation, epistemology, social implications, and political outcomes.

The conceptual engagements with GIS that are developed in this third section return us to the multiplicities and hybridities that are central to qualitative GIS. The contributing authors in Section 3 treat GIS as a collection of analysis and representation techniques, as a set of research practices that are embedded in multiple disciplinary accounts of and struggles over knowledge making in research, and as a set of social and political practices that engage with the broader world. They show that GIS might be both studied and practiced from multiple epistemological perspectives, even within a single project. They point to multiple nodes where the 'qualitative' in qualitative GIS might be activated – in the socio-political or institutional contexts of GIS creation and use, the knowledge creation processes in which it is embedded, and the modes of analysis it is used to create. Finally, they demonstrate a hallmark of qualitative GIS that threads through all of the chapters in this collection: its persistent critical reflection upon the contributions and silences of different ways of knowing, and the social and political power of different forms of representation and analysis.

NOTES

1 Pavlovskaya's and Wilson's chapters in this volume develop in much more detail how qualitative GIS has emerged from these 'GIS and society' critiques, and how it may be situated within the new forms of GIS research and practice that have responded to them.

2 Of course, new developments such as online mapping and spatial services or wireless GPS enabled handheld devices are altering this understanding of what constitutes a GIS, and many of these new technologies offer exciting possibilities for integrating multiple qualitative representations of geographic knowledge into digital technologies. For more detailed discussion, see Miller (2006), Sheppard (2006), Goodchild (2007), and Elwood (2009b).

3 It is not only textual or linguistic forms of evidence that can be qualitative by virtue of containing negotiated interpretive meanings. A photograph, map, or other visualization may be qualitative by virtue of the interpretations of place produced by its creator, and those meanings may be advanced as users of the image continue to interpret and use it themselves.

REFERENCES

Al-Kodmany, K. (2002) 'GIS and the artist: shaping the image of a neighborhood through participatory environmental design', in W. Craig, T. Harris and D. Weiner (eds), *Community Participation and Geographic Information Systems*. London: Taylor and Francis. pp. 321–9.

Chrisman, N. (2002) *Exploring Geographic Information Systems*. New York: Wiley.

Cieri, M. (2003) 'Between being and looking: queer tourism promotion and lesbian social space in greater Philadelphia', *ACME: An International E-Journal for Critical Geographies*, 2 (2): 147–66.

Crang, M. (2005) 'Qualitative methods: there is nothing outside the text?', *Progress in Human Geography*, 29 (2): 225–33.

Cresswell, T. (2003) *Research Design: Qualitative, Quantitative and Mixed Methods Approaches*, 2nd edn. Thousand Oaks, CA: Sage.

Dennis, S. (2006) 'Prospects for qualitative GIS at the intersection of youth development and participatory urban planning', *Environment and Planning A*, 38 (11): 2039–54.

Elwood, S. (2006) 'Beyond cooptation or resistance: urban spatial politics, community organizations, and GIS-based spatial narratives', *Annals of the Association of American Geographers*, 96 (2): 323–41.

Elwood, S. (2009a) 'Mixed methods: thinking, doing, and asking in multiple ways', in D. DeLyser, M. Crang, L. McDowell, S. Aitken and S. Herbert (eds), *The Handbook of Qualitative Research in Human Geography*. London: Sage.

Elwood, S. (2009b) 'Geographic information science: new geovisualization technologies – emerging questions and linkages with GIScience research', *Progress in Human Geography*, 33 (3).

England, K. (1993) 'Suburban pink collar ghettos: the spatial entrapment of women?', *Annals of the Association of American Geographers*, 83 (2): 225–42.

Goodchild, M. (2007) 'Citizens as sensors: the world of volunteered geography', *GeoJournal*, 69 (4): 211–21.

Harris, T. and Weiner, D. (1998) 'Empowerment, marginalization, and community-oriented GIS', *Cartography and Geographic Information Systems*, 25 (2): 67–76.

Jiang, H. (2003) 'Stories remote sensing images can tell: integrating remote sensing analysis with ethnographic research in the study of cultural landscapes', *Human Ecology*, 31 (2): 215–32.

Knigge, L. and Cope, M. (2006) 'Grounded visualization: integrating the analysis of qualitative and quantitative data through grounded theory and visualization', *Environment and Planning A*, 38 (11): 2021–37.

Kwan, M. (2002) 'Feminist visualization: re-envisioning GIS as a method in feminist geography research', *Annals of the Association of American Geographers*, 92 (4): 645–61.

Kwan, M. (2007) 'Affecting geospatial technologies: toward a feminist politics of emotion', *The Professional Geographer*, 59 (1): 22–34.

Kwan, M. and Knigge, L. (2006) 'Doing qualitative research with GIS: an oxymoronic endeavor?', *Environment and Planning A*, 38 (11): 1999–2002.

Kwan, M. and Lee, J. (2004) 'Geovisualization of human activity patterns using 3D GIS: a time-geographic approach', in M. Goodchild and D. Janelle (eds), *Spatially Integrated Social Science*. New York: Oxford University Press. pp. 48–66.

Lake, R. (1993) 'Planning and applied geography: positivism, ethics, and geographic information systems', *Progress in Human Geography*, 17 (3): 404–13.

Lawson, V. (1995) 'The politics of difference: examining the quantitative/qualitative dualism in poststructural feminist research', *The Professional Geographer*, 47 (4): 449–57.

Matthews, S., Detwiler, J. and Burton, L. (2005) 'Geo-ethnography: coupling geographic information analysis techniques with ethnographic methods in urban research', *Cartographica*, 40 (4): 75–90.

Mattingly, D. and Falconer Al-Hindi, K. (1995) 'Should women count? A context for the debate', *The Professional Geographer*, 47 (4): 427–35.

Maxcy, S. (2003) 'Pragmatic threads in mixed methods research in the social sciences: the search for multiple modes of inquiry and the end of the philosophy of formalism', in A. Tashakkori and C. Teddlie (eds), *Handbook of Mixed Methods in Social and Behavioral Research*. Thousand Oaks, CA: Sage. pp. 51–89.

McCann, E. (2008) 'Expertise, truth, and urban policy mobilities: global circuits of knowledge in the development of Vancouver, Canada's "four pillar" drug strategy', *Environment and Planning A*, 40 (4): 885–904.

McLafferty, S. (2002) 'Mapping women's worlds: knowledge, power and the bounds of GIS', *Gender, Place and Culture*, 9 (3): 263–9.

Miller, C. (2006) 'A beast in the field: the Google Maps mashup as GIS', *Cartographica*, 41 (3): 187–99.

Moss, P. (1995) 'Embeddedness in practice, numbers in context: the politics of knowing and doing', *The Professional Geographer*, 47 (4): 442–9.

Moss, P. (ed.) (2003) *Feminist Geography in Practice: Research and Methods*. London: Blackwell.

Nast, H. (1994) 'Women in the field: critical feminist methodologies and theoretical perspectives', *The Professional Geographer*, 46 (1): 54–66.

Nightingale, A. (2003) 'A feminist in the forest: situated knowledges and mixing methods in natural resource management', *ACME: An International E-Journal for Critical Geographies*, 2 (1): 77–90.

Pain, R., MacFarlane, R., Turner, K. and Gill, K. (2006) 'When, where, if and but: residents qualify the effect of streetlighting on crime and fear in their neighbourhoods', *Environment and Planning A*, 38 (11): 2055–74.

Pavlovskaya, M. (2002) 'Mapping urban change and changing GIS: other views of economic restructuring', *Gender, Place and Culture*, 9 (3): 281–9.

Pavlovskaya, M. (2006) 'Theorizing with GIS: a tool for critical geographies?', *Environment and Planning A*, 38 (11): 2003–20.

Pickles, J. (1995) 'Representations in an electronic age: geography, GIS, and democracy', in J. Pickles (ed.), *Ground Truth: The Social Implications of Geographic Information Systems*. New York: Guilford. pp. 1–30.

Robbins, P. (2003) 'Beyond ground truth: GIS and the environmental knowledge of herders, professional foresters and other traditional communities', *Human Ecology*, 31 (2): 233–53.

Rocheleau, D. (1995) 'Maps, numbers, text, and context: mixing methods in feminist political ecology', *The Professional Geographer*, 46 (1): 458–66.

Schuurman, N. (2000) 'Critical GIS: theorizing an emerging discipline', *Cartographica*, Monograph 53.

Schuurman, N. (2006) 'Formalization matters: critical GIS and ontology research', *Annals of the Association of American Geographers*, 96 (4): 726–39.

Sheppard, E. (1995) 'GIS and society: towards a research agenda', *Cartography and Geographic Information Systems*, 22 (1): 5–16.

Sheppard, E. (2006) 'Knowledge production through critical GIS: genealogy and prospects', *Cartographica*, 40 (4): 5–21.

Shiffer, M. (2002) 'Spatial multimedia representations to support community participation', in W. Craig, T. Harris and D. Weiner (eds), *Community Participation and Geographic Information Systems*. London: Taylor and Francis. pp. 309–20.

Tashakkori, A. and Teddlie, C. (2003) 'The past and future of mixed methods research: from triangulation to mixed model design', in A. Tashakkori and C. Teddlie (eds), *Handbook of Mixed Methods in Social and Behavioral Research*. Thousand Oaks, CA: Sage. pp. 671–701.

Weiner, D. and Harris, T. (2003) 'Community-integrated GIS for land reform in South Africa', *The URISA Journal*, 15: 61–73.

2

NON-QUANTITATIVE GIS

Marianna Pavlovskaya

ABSTRACT

Despite its relatively weak quantitative functionality, GIS is primarily associated with statistical and quantitative spatial analysis. This creates a particular representation of GIS as linked to traditional understandings of science and technology and, critically, to corporate power and institutions of control. In addition, constructing GIS as solely quantitative prevents it from being used for qualitative analysis, non-quantitative spatial analysis, and progressive research that often (although not always) relies upon non-quantitative research methods. GIS is, however, well suited for particular forms of qualitative research. For example, it allows for a rich visualization of information in the form of maps and other types of graphic data representation. In this sense, cutting-edge research in geovisualization is directly supporting non-quantitative uses of GIS. In addition to geovisualization, other recent research illustrates not only that a qualitative GIS is possible and growing but that it fulfills an important epistemological function. This function consists of the ability to visualize and investigate social phenomena that cannot be represented by quantitative databases (whether governmental, commercial, or user created) or analyzed by traditional quantitative and statistical techniques. Not only does qualitative GIS contribute to furthering our scientific understanding of the world by expanding the range of usable epistemologies, but it also supports research agendas that are committed to progressive politics and challenge the status quo. Finally, qualitative GIS also contributes to advances in social theory because it easily incorporates space into our thinking about the world and allows us to ask research questions that can only be addressed through mixed methods research.

INTRODUCTION

Just a few years ago, critical GIS (geographic information systems or science) scholars had to argue that a qualitative GIS was even possible and that it could contribute to a valid and robust research methodology (Bell and Reed, 2004; Knigge and Cope, 2006; Kwan, 2002a; Kwan and Knigge, 2006; Matthews et al., 2005; Pavlovskaya, 2002). Today, we are contributing to a textbook on qualitative GIS written for a wide audience of students, academics, and GIS practitioners.[1] This remarkable development is related to and enhanced by the recent powerful re-entry of qualitative and ethnographic methods into human geography after a period of relative undervaluing of the humanistic tradition. Major recent methodological texts now include thorough discussions of qualitative research (e.g. Babbie, 2000; Clifford and Valentine, 2003; Cloke et al., 2004; Tashakkori and Teddlie, 2002; see Crang, 2002 for an overview) and the politics of doing such research today are widely debated (Crang, 2002; St Martin and Pavlovskaya, forthcoming b).

A recent emphasis on mixed methods research has also contributed to the emergence of qualitative GIS. Previously opposed to quantitative methods, critical human geographers have re-envisioned their use in conjunction with qualitative modes of explanation (Elwood, 2006a; Kwan, 2002a; 2002b; 2002c; Lawson, 1995; McLafferty, 1995; 2002; Sheppard, 2001; St Martin, forthcoming). Similarly, and also a result of the growing availability of digital spatial data and user-friendly software for their viewing (e.g. Google Earth), geospatial technologies are increasingly incorporated into mixed methods approaches. Combining GIS with qualitative methods allows critical human geographers to use the analytical and representational power of GIS as well as to get around its limitations with respect to certain forms of analysis (see Introduction, this volume).

Qualitative GIS has also made relevant to GIS research the debates in critical human geography about the political nature of production of knowledge and representation initiated by feminist and poststructuralist critics of science (Foucault, 1980; Gibson-Graham, 2000; Haraway, 1991; Katz, 2001; Rose, 1992). Its effects are felt throughout the whole process, from defining research problems and choosing methods to producing findings and interacting with research participants, assistants, and colleagues. In the words of Cindi Katz (1992), knowledge production 'oozes with power'. This could not be more important than in the case of GIS which is, at once, a powerful research and representational tool, a charismatic technology, and a multi-billion-dollar industry. There is also a powerful narrative about 'what GIS is' that defines what it can or should and cannot or should not do (St Martin and Wing, 2007). Therefore, GIS practice and scholarship also result in silencing certain research practices and uses that do not fit these definitions.

This chapter argues for qualitative GIS as a powerful research strategy by exposing some of the silences that are produced by the prevailing narrative of 'GIS as a quantitative tool'. While this narrative grants irrefutable scientific authority to GIS, it also silences its non-quantitative functionality that, I argue, actually constitutes its core in many respects. Breaking silences around the affinity of GIS with *qualitative* analysis opens it up to ethnographic and mixed methods research. The chapter begins by examining GIS as a power relation negotiated in broader epistemological struggles within geography. It then proceeds to deconstruct the prevailing notion of GIS as a quantitative tool and highlight its capabilities for qualitative research, including rich functionality for visualization. Lastly, I use examples from recent research that illustrate that qualitative GIS not only is possible but can also fulfill an important epistemological function that quantitative research cannot.

GIS AS POWER RELATION

GIS indeed represents power to most audiences: it stands for funding and research grants, jobs, information, student enrollments, mesmerizing images on the computer screen, best solutions and locations, and the power to convince. This power derives from the position of GIS at the intersection of science, technology, and visuality. First, GIS is firmly associated with quantitative analysis and the scientific method. Second, its flesh and blood are computers and digital information. And, third, it expresses the very fascination of Western science and geography with vision, seeing, and looking as a primary and supposedly objective way of knowing, which is in fact partial, embodied, and masculinist (Cosgrove and Daniels, 1988; Haraway, 1991; Rose, 1992; Sui, 2000). Similar to cartography (Crampton, 2001), the power of GIS lies in its ability to create visual images of the world based on scientific information, to unveil previously hidden natural and social landscapes with an authority of science. The prevailing image of GIS as a powerful juncture of science, technology, and authority serving big business and the government has been created and sustained by many actors. These include academic departments where GIS is taught, corporations where the technology is developed, and various groups of users from grassroots organizations to businesses and governments worldwide (Kwan, 2002a; Longley et al., 2005; Schuurman and Pratt, 2002; St Martin and Wing, 2007). As a representational tool and a socially embedded technology, GIS is indeed 'oozing with power'.

Not surprisingly, then, GIS has been passionately debated in geography since the early 1990s (see Schuurman, 1999; 2000 for details). These debates concern not only the field of GIS *per se* but also geography's identity as a discipline (Goodchild, 1991; Kwan, 2002a; Openshaw, 1991; 1998; Sui and Morill, 2004), practices of knowledge production and representation (Bell and Reed, 2004; Crampton, 2001; Elwood, 2006a; 2006b; Elwood and Martin, 2000; McLafferty, 2002; 2005; Pavlovskaya and St Martin, 2007; Sheppard, 2005), and the relationships between GIS and economic and social power (Crampton, 2003; Pickles, 1995; 2004; Smith, 1992). In other words, the debates about GIS have been intimately related to epistemological struggles over scientific authority. It is a power relation negotiated by different practices of knowledge production in human geography identified with quantitative and qualitative methods. This understanding helps to explain the passion surrounding GIS, its continued transformations, and even its integration with qualitative methods, the last of which was recently unthinkable but now forms the subject of this book.

Historically, the field of GIS has been associated with quantitative spatial science in geography. Seen as socially and scientifically progressive in the 1950s and 1960s, since the 1970s this tradition has been critiqued by Marxists, by feminists, and later by poststructuralists for scientific and social conservatism. The scientific conservatism resulted from a positivist epistemology while the social conservatism of mainstream social science stemmed from its general support for the economic and social institutions of capitalism, which the new approaches sought to examine critically. It became unthinkable to practice progressive social science while assuming objectivity, a value-free researcher, and clear separation of the subject and the object of the research. Researchers concerned with class, gender, sexuality, and race denounced not only the conservative politics but also the methodologies that were linked to the production of such scholarship, of which quantitative analysis was a major tool. Critical scholars instead turned to the qualitative methods of humanistic geography and recast them within critical geography paradigms (Cloke et al., 1991; Katz, 2001; Livingstone, 1992; Staeheli and Mitchell, 2005). At some points in the philosophy and methods debates of the late twentieth century, choosing a method (e.g. regression analysis or ethnography) represented a choice of not only one's philosophy of science but also one's professional and personal politics.

GIS entered geography in the midst of these debates. It was largely constituted by these debates in a specific way – as a quantitative tool of spatial science. In various texts, the language of GIS is that of science, measurement, spatial data models, spatial analysis, sampling, geocomputation, calculation, databases, data transformation, validation and so on (examples are Clarke et al., 2002; Crisman, 2002; de Smith et al., 2007; Longley et al., 2005). The landscape of the GIS community today is very complex but a number of authors have shown that within it both 'quantitative' proponents and 'qualitative' critics of GIS contributed to this image of the field (Schuurman, 2000; Schuurman and Pratt, 2002; St Martin and Wing, 2007). For the proponents, the connection of GIS to science, quantitative geography, spatial analysis, and computerization validated its growth and has been a source of pride (Clarke, 1999; Goodchild, 1991; Longley et al., 2005). Today, many professors and students equate GIS with geography and see it as a scientific solution to most geographic problems and the most important job skill for graduating students (Flowerdew, 1998; Longley et al., 2005; Openshaw, 1998; Sui and Morill, 2004). Part of this valorization is represented by the shift to the term GIScience (GISc), which has replaced the more mundane term GISystems (Wright et al., 1997), implying a transition from simply a tool to a theory of digital representation of the world and its analysis.

The critics, too, linked GIS to spatial science and quantitative geography. In contrast to GIS academics, however, they focused on the epistemological and social conservatism embedded in its representational, technological, and scientific authority. For many of them, GIS was a problem, not a solution. In their view, GIS reduces places and people to digital 'dots' and enables those in power to make decisions without involving local communities. GIS serves corporate profit making and state interests; facilitates surveillance, control, and warfare; masks social and economic inequality; supports the seeming objectivity of data and analysis; perpetuates a male-dominated field; serves as a successor to imperial cartography; and is an essentially undemocratic information technology due to its high cost, unequal access, and need for expert knowledge (Armstrong and Ruggles, 2005; Crampton, 2003; Curry, 1997; Dobson and Fisher, 2003; Goss, 1995; Kwan, 2002c; Pickles, 1995; 1997; Rocheleau, 1995; Schuurman, 2002; Sheppard, 1993; Smith, 1992; Treves, 2005). Together, all these aspects of GIS practice left no room for its application within Marxist, feminist, poststructural, and postcolonial frameworks. Seen as solely quantitative and technocratic, GIS was overwhelmingly denounced by critical human geographers in the 1990s. In short, despite the normative disagreements of those involved in the debates, GIS emerged as a singular tool to be used within a particular practice of knowledge production (Kwan, 2002a; Schuurman and Pratt, 2002; St Martin and Wing, 2007), a seductive technology firmly linked to quantitative science, power, and capital.

And yet, the situation has begun to change in the past decade or so. A body of knowledge loosely defined as 'critical GIS' has emerged that has enabled innovative, non-quantitative, and progressive uses of and perspectives on GIS (for overview, see Schuurman, 2002; Sheppard, 2005). In some ways, critical GIS is a result of the growing theoretical pluralism of the past three decades. Partiality of knowledge has become an acceptable epistemological stance that necessitates conversation and makes explicit one's responsibility for knowledge production practices. Feminist scholarship, in particular, has transformed social sciences by bringing in the excluded subjects and thoroughly changing research ethics. These developments encouraged GIS scholars to think about the possibilities of GIS within diverse theoretical frameworks.

The assumed vast differences in scientific rigor between quantitative and qualitative methods have also been profoundly rethought on both the qualitative and quantitative sides (see, for example, Baxter and Eyles, 1997; Cloke et al., 2004; Crang, 2002; Poon, 2004). Qualitative research is no longer considered to be a precursor or an afterthought of a large-scale quantitative study, equal in significance to 'coffee-table talk' (Openshaw, 1998). Both approaches today are seen as different but equally powerful research strategies if used appropriately. While one focuses on the power of generalization and statistical representation, the other enables explanation, understanding, and theoretical representation (Strauss, 1995). Both, however, are socially embedded practices and, therefore, can be logical or irrational (Barnes, 2001), sophisticated or simple, large or small in terms of amount of data, spatial scale, time, and labor, as well as sloppy or rigorous. With qualitative methods regaining their authority, geographers began to 'discover' qualitative aspects even in the established quantitative research tools such as GIS – as discussed later in the chapter (see also Elwood, 2006a; Knigge and Cope, 2006; Kwan and Knigge, 2006; Pavlovskaya, 2006).

But in addition to the above, it is the ongoing delinking of epistemology and methods across social sciences that has enabled innovative and non-quantitative GIS practices. The assumed alignments of ontology, epistemology, and methods within a particular paradigm (e.g. spatial science with quantitative methods and feminism with qualitative methods) have been destabilized, and both types of methods are increasingly practiced across different epistemological frameworks. Feminist, Marxist, and poststructuralist geographers found ways to incorporate quantitative analysis (see Hanson, 1997; Lawson, 1995; McLafferty, 1995; Plummer and Sheppard, 2001; Sheppard, 2001) and the strictly quantitative scholars have begun to appreciate qualitative reasoning (Poon, 2004). Today, 'quantitative' no longer stands for 'positivist' even among social theorists (but see Amin and Thrift, 2000) and 'qualitative' no longer means lack of science. The choice of methods became more pragmatic but no less rigorous because the responsibility of researchers for their choices has been made explicit. It is the internal consistency, transparency, and reflexivity of the methods, their ability to acquire and analyze needed information, either quantitative or qualitative, that have become most important. In this context, we can decouple GIS, too, from its assumed epistemological home and imagine its valid uses in other research frameworks.

The related rise of mixed methods in social sciences and geography (Creswell, 2003; Tashakkori and Teddlie, 2002) also opens GIS to new imaginations. In mixed methods projects, researchers use quantitative and qualitative methods either sequentially at different stages or interactively at all stages (Knigge and Cope, 2006). They combine methods to cross-reference and triangulate data but also to consider incongruencies in data as research opportunities. Geographers in particular are increasingly keen to combine different methods with GIS when research goals make it appropriate (Kwan, 2007; McLafferty, 2005; Nightingale, 2003; Pavlovskaya, 2004; Robbins and Maddock, 2000; Sheppard, 2005; St Martin, forthcoming). Finally, the so-called 'spatial turn' in social sciences and humanities has increased attention to the spatiality of human experiences and encouraged thinking about space in non-quantitative and visual terms. The language of boundaries, flows, and territories, as well as that of cartography and maps, has found its way into wider social research and art. Not surprisingly, GIS is now used outside its traditional technical fields and is being rapidly integrated with the latest multimedia and web-based technologies (Peng, 2001; Pavlovskaya, forthcoming).

With all these developments in place, it is vital to articulate GIS as a strategy for mixed methods research that transgresses the established epistemological boundaries. While important work in this direction has already begun (Craig et al., 2002; Elwood, 2006a; 2006b; Elwood and Leitner, 1998; Knigge and Cope, 2006; Kwan, 2002a; 2002b; 2002c; 2007; McLafferty, 2005; Pavlovskaya, 2002; Schroeder, 1996; Schuurman and Pratt, 2002; Sheppard, 2005; Sieber, 2004), it would be too soon to say that GIS has seamlessly joined the diverse practices of knowledge production. The dominant discourse of GIS remains that of a quantitative tool; it tends to alienate and marginalize other research methods; its corporate, military, and state applications prevail; and the industry itself is increasingly dominated by a single corporate developer. Given the representational power of GIS and its rapidly spreading applications, reclaiming geospatial technologies for critical geographies, qualitative research, and progressive politics has been at no time more crucial than it is now.

OPENINGS FOR NON-QUANTITATIVE GIS

Thinking of GIS as a negotiated power relation in the production of knowledge instead of a given, unchangeable technique helps to see GIS as "constantly remade through the politics of its use, critical histories of it and the interrogation of concepts that underlay its design, data definition, collection and analysis. In other words, futures of GIS are contested and openings exist for new meanings, uses and effects" (Pavlovskaya, 2004; 2006). Below I offer one strategy for enabling new meanings and uses of GIS. In particular, I refocus the prevailing narrative of GIS that constructs it as a quantitative technology on to commonly overlooked and, therefore, silenced non-quantitative functionality. I do so by identifying a series of openings or contradictions in GIS practice that break silences and produce possibilities for qualitative GIS. They show that GIS has much greater affinity with qualitative research than we commonly think.

Opening 1: GIS origins are mainly non-quantitative

To begin, the very origins of GIS are mainly non-quantitative. It evolved from a variety of fields besides quantitative geography and combines diverse bodies of knowledge. They include geography (mapping and spatial analysis), computer science (automation and computing), land use planning and census administration (handling and display of large databases), remote sensing (image processing and land cover analysis), and geodesy and the military (spatial accuracy and georeferencing) (Clarke, 1999; Flowerdew, 1998). In other words, using GIS requires specialized knowledge but this knowledge is different from the expertise in quantitative analysis.

Opening 2: computerization is not quantification

Since their early days, computer technologies have represented science. The beginnings of quantitative geography in the 1950s coincided with and were facilitated by the introduction of computers, and, as an emerging field, spatial science benefited from this association (Barnes, 2000; 2001). Computers created an illusion of accuracy in data and calculation, handled large amounts of information, and, just like scientific data, needed systematically organized datasets. GIS, too, handles large and structured databases, offers specific analytical tools, and is part of expanding information technologies. Yet, modern computing supports a whole range of non-analytical and non-quantitative activities (e.g. paying bills or playing games). Researchers, too, use a broad range of software packages, many of which automate non-analytical tasks (e.g. word processing or bibliographic software) or non-quantitative analysis (e.g. graphic display or qualitative data analysis using ATLAS.ti).

In the early days of computers, GIS programmers were academics who developed software to automate their spatial analyses. The link between GIS and scientific modeling was prominent (Schuurman, 1999). Today, with rare exceptions (e.g. IDRISI), the development of mass-consumption GIS software is in the hands not of academics but of corporations. Creating any computer application requires programming skills but few applications require quantitative spatial analysis. Moreover, most existing spatial analysis algorithms predate computerization and were incorporated into the software long after they were developed. Thus, the two bodies of knowledge – programming and quantitative analysis – are separate types of expertise (see also Crisman's, 2002: iii comment about their conflation).

Furthermore, most of the diverse functionality of GIS (e.g. data visualization and querying, overlays, etc.) is made efficient by automation but remains quantitatively and statistically unsophisticated. Remote sensing software is far more quantitative in this sense because even basic image classification techniques use complex statistical procedures (e.g. cluster analysis, maximum likelihood classification, principal component analysis, etc.), as do many non-spatial statistical software applications (e.g. SPSS or Statistica). Thus, computerization enabled GIS to process digital information but in itself it did not make this information processing more quantitative.

Opening 3: spatial analysis in GIS is non-quantitative

Surprisingly, only a modest share of GIS functionality involves quantitative spatial analysis (Eastman, 2003; Flowerdew, 1998; Openshaw, 1998; Schuurmann, 1999). Even popular GIS textbooks admit that "most GIS packages have contained only rudimentary tools for spatial analysis" (Clarke, 1999: 181). Most GIS users, therefore, have access to only basic techniques such as overlay, linear distance calculations, buffering, determining neighbors, or summarizing data within new geographic boundaries. While very illuminating, these techniques do not require knowledge of advanced mathematics from GIS users. Examples include calculating employment opportunities within a certain distance of women's homes (Hanson et al., 1997), overlaying locations of banks engaged in predatory lending with census data to reveal their target populations (Graves, 2003), and mapping hazardous accident sites by census units to calculate exposed populations (Margai, 2001). In truth, most spatial techniques available in GIS require spatial imagination (e.g. to grasp buffering or overlay), logical thinking (e.g. combining layers in site selection or multi-criteria evaluation), or intuitive grasp (in visual examination) and, therefore, replicate qualitative reasoning common to all geographic research. This affinity with human reasoning has been also obscured for a long time by the unfriendly user interfaces of many GIS programs and applications.

Ironically, much of the recent GIS research seeks to enhance precisely these qualitative aspects. Fuzzy sets theory, artificial intelligence, cellular automata, chaos and complexity theory, agent-based modeling, and Bayesian probability all attempt to model human reasoning that involves multiple connections, blurred ontological categories, uncertainty in decision making, and the pragmatic use of partial knowledge (Ahlqvist, 2004; Openshaw, 1998; Sheppard, 2001). Ironically, as is the case with buffering, for example, the challenge is not in the mathematical sophistication of the technique itself – it is quite non-quantitative – but in designing and mathematically implementing an algorithm that replicates human reasoning (e.g. decisions made under uncertainty) or a conceptually simple spatial operation (e.g. buffering).

Finally, much of GIS literature deals with such methodological issues as ecological fallacy and modifiable areal unit problems (Openshaw and Taylor, 1979; Wong, 2003; 2004), questions of appropriate spatial resolution and locational accuracy (Scott et al., 1997), methods for distance calculations (Wang, 2000), representation of objects as either continuous or discrete, the ontological structure of objects (Fonseca et al., 2000), and so on. While there are GIS-specific tasks such as digital spatial data models (e.g. Ahlqvist, 2004), other issues, again, are common to geographic analysis in general. Matters of conceptualization, they are not in themselves quantitative problems. In the end, despite the consistent labeling of GIS as a quantitative tool, its most commonly used functions are rather qualitative.

Opening 4: digital data are not always for counting

Digital data representation, including GIS databases (spatial and attribute), is usually associated with large numerical datasets, but upon closer examination it also has little 'quantitative' content in itself. All information for computer use must be represented digitally and, therefore, appropriately coded. This means that digital data have embedded histories; they are not neutral descriptors of the world but social constructs, that is they are products of those who created them, their purpose, and their approach. Furthermore, digital data must be coded regardless of whether they are quantitative or qualitative and whether they are to be analyzed quantitatively or not. In word processors, too, letters are expressed with binary code but are not for use in a regression model. They are digital because they cannot be stored and visualized in the computer otherwise. While coding already implies categorization, fixity, and structured ontology (Dixon and Jones III, 1996; Doel, 2001; Jones III and Dixon, 1998; Lawson, 1995), using numbers to express qualitative properties of geographic objects does not yet amount to quantitative analysis. For example, topology lies at the heart of vector models and represents very structured but non-quantitative spatial relationships. Digital data like photographs or sound are non-quantitative too. In short, digital representation does not substitute for quantitative analysis.

Opening 5: database management and querying are based upon geographic location

Suggesting its origins in an empiricist scientific tradition, GIS easily handles large amounts of data (Flowerdew, 1998). Compared to non-spatial database management systems, it organizes data in a unique way – by geographic location. Assembling and structuring spatial 'facts' in a geographic database (e.g. land parcel, TRI, or census data) allow for versatile querying and display of datasets comprising thousands of spatial units and variables describing them. Spatial databases also allow for unique merging of information from different sources. As digital information and especially spatially referenced data continue to explode, the role of GIS in meaningfully organizing these datasets will only increase (St Martin and Pavlovskaya, forthcoming a; Sui and Morrill, 2004). This extraordinary ability of GIS to manage and query spatial data is, however, conflated with an ability to quantitatively analyze them.

Database development and maintenance – tasks that consume enormous amounts of time, as GIS textbooks frankly acknowledge (Clarke, 1999) – do not involve quantitative analysis at all. Digitizing and cleaning spatial layers (e.g. snapping nodes, building polygons, or georeferencing satellite data), merging spatial databases, as well as entering, organizing, and verifying attribute data, do require knowledge of geodesy, geometry, data structures, and the subject matter of the database but do not require knowledge of advanced spatial analysis or modeling. Building a GIS database for a qualitative project would require the same technical skills and expertise as for a quantitative project (see Jung, this volume, for an example).

Digital attribute data themselves, too, are often qualitative and include names (e.g. of owners of land parcels, businesses, or street addresses) or types (e.g. of roads, settlements, soils, or polluting facilities). While not suitable for quantitative analysis, such data can, however, be queried and logically manipulated using SQL (structured query language) in order to find geographic features with particular characteristics. Even complex attribute and spatial queries, however, require logical thinking and a spatial imagination rather than statistical or mathematical expertise. SQL also enables numerical manipulation, but advanced calculations are less common in a GIS and, as we will see, are most often performed in a non-GIS environment.

Opening 6: mathematical modeling and statistics are still outside GIS

Even more revealing, the advanced GIS and quantitative geographers seldom use commercial GIS for analysis. They often need specific algorithms that are absent in commercial packages or have their details concealed. GI scientists program their own spatial analytical routines and display the results in the existing GIS software (Flowerdew, 1998; Kwan, 1999a; 1999b; Openshaw, 1998). Further, the community of quantitative geographers is quite different from the GIScience community (Fotheringham, 1997; Goodchild, 1991; Poon, 2004). They publish in different journals (Miller and Wentz, 2003) and use non-GIS quantitative analysis packages (e.g. MatLab, IDL, SPSS, Statistica, or MS Access or Excel) or existing specialized models (e.g. for atmospheric circulation, plume dispersion, or crime 'hot spot' identification), or write programs themselves (e.g. the geographically weighted regression (GWR) software developed by Fotheringham et al., 2002). This is true even for studies that explicitly focus on spatial processes (e.g. Margai, 2001; Plummer, 2000; Poon, 2004). While these routines may eventually become add-ons to GIS, the point is that they are not among the most widely used or initially available GIS functions.

The fact that many statistical techniques including regression analysis simply cannot be applied to spatial data (Getis and Ord, 1996) also limits the quantitative capacity of GIS. For example, proximity generates autocorrelation in spatial distributions and this violates fundamental assumptions of data independence in conventional statistics. Initially developed for non-spatial data, these statistics were imported into geography without proper adjustment (Barnes, 1998; 2001; Flowerdew, 1998; Sheppard, 2001). Thus, ignoring locational information, unfortunately, cancels out the very difference GIS could have made.

Most available statistics, even spatial statistics, also calculate parameters (e.g. autocorrelation coefficients or regression equations) that apply to the entire study region and ignore local variation in their values. This defeats the purpose of geographic analysis and leads to creation of mis-specified or poorly fitting models (Fotheringham, 1997; Fotheringham et al., 2002). In addition, the available methods do not do well in modeling dynamic processes, incorporating individual-level data (Miller, 2003), or representing interactions across geographic scales and networks (Poon, 2004). Only recently have geographers developed advanced geostatistical methods that address these and other problems of spatial modeling (Barnes, 1998; Fotheringham, 1997; Getis and Ord, 1996; Poon, 2004; Sheppard, 2001). These techniques, however, usually are available in software packages separate from GIS or only recently incorporated. Visualizing spatial distributions remains the main functionality that quantitative modelers seek and use in GIS.

Opening 7: visualization can be a qualitative analytical technique

In the end, visualization is arguably the most powerful and widely used function in GIS. Like other tools for graphic data display, GIS makes spatial information immediately accessible to our minds. Scholars prefer to 'see' the data, either quantitative or qualitative, in order to assess their quality, suitability, or completeness, and to 'see' the results in order to decide whether each transformation or query is correct or not. Even in purely quantitative research mapping, value distributions help, for example, identify model mis-specification problems (Fotheringham, 1997). Visual examination itself does not involve mathematical calculation, but is a powerful analytical technique.

More importantly, however, visualization is the source of the seductive rhetoric of GIS, the rhetoric that combines the power of maps with the power of science and technology. Maps communicate spatial information in a particularly synergistic way. Far from simply conveying data, maps convey power because they express the authority to selectively represent people and places (Crampton, 2001; Edney, 1997; Harley, 1992; Lewis, 1998; Sparke, 1998; St Martin, 1995). Placing this power into the realm of information technology, GIS further validates maps as scientific constructs (Lake, 1993; Sheppard, 1993). GIS unveils worlds to researchers, policy makers and the public, worlds made 'true' by the assumed legitimacy of data and visual displays.

Not surprisingly, the GIS industry has always focused on display functions as a way to analyze data as well as conquer hearts. GIS academics, too, have produced vast research on visualization including its technical, computer-related, methodological, cognitive, and social theoretical issues (Knigge and Cope, 2006; Kwan, 2002a; MacEachren, 1994). The recent surge in work on geovisualization and exploratory spatial data analysis (ESDA) in particular demonstrates that visualization is no longer a means to represent analytical results but a means of analysis itself. In the past, cartography served to communicate the results of research to the public as suggested by the map communication model (Robinson and Petchenik, 1976). In this model, the cartographer's task was to best communicate (by properly choosing symbols, colors, themes, scale, etc.) the already derived scientific knowledge to the public who were to passively receive that knowledge. In the past decade, however, Alan MacEachren (1994; MacEachren et al., 1999) has advanced the concept of visualization as an analytical tool linked to an automated data display in GIS. Here, the research process itself becomes a focus. Assisted by computerized visualization, a researcher or a GIS user interactively and iteratively analyzes the data and immediately displays the results in a number of ways. She or he explores both the data and the analytical techniques and, by directly interacting with a GIS, becomes simultaneously the author and the reader of the map (Crampton, 2001). The GIS-based map is transformed from a vehicle for delivering knowledge into an interactive knowledge production practice including the potential to become the primary medium for communication between scientists themselves (MacEachren et al., 2004). The potential of integrating GIS visualization with qualitative analysis is particularly promising (see 'grounded visualization' presented in Knigge and Cope, 2006; also Knigge and Cope, this volume).

Visualization is powerful because it provides opportunities for heuristic (non-logical) understanding of data and processes. While an important component of human decision making, this understanding cannot be achieved by rational analysis but complements it. The visual impact of GIS also depends upon emotions and other irrational sentiments (Kwan, 2002a; 2007) that run counter to the dry logic of quantification. In short, visualization is the most telling non-quantitative functionality of GIS.

To conclude, the most widely used functions in GIS, such as visualization, database development, management, and querying, are not at all quantitative despite that the dominant narratives construct GIS as a quantitative analytical tool. The alternative reading presented above highlights the limited use of GIS in quantitative analysis and points to its unacknowledged potential for qualitative research that I will turn to now.

THEORIZING WITH GIS

The possibilities of a distinctly qualitative GIS within critical human geographic research have been open up by critical GIS scholars. Public participation GIS (PPGIS) scholars, for example, have long been working on making GIS and other geospatial technologies, including internet-based geographic information, more democratic. They seek to empower marginalized groups through the use of these technologies (Craig et al., 2002; Elwood, 2006b; Gilbert and Masucci, 2004). Feminist geographers, however, were among the first to argue against essentializing GIS as a positivist and masculinist technology and for using it in feminist research (Hanson, 2002; Kwan, 2002a; 2004a; McLafferty, 2002; 2005; Pavlovskaya, 2002; Schuurman, 2002; Schuurman and Pratt, 2002). We are witnessing the emergence of a new mapping subject who is a male or female GIS researcher/user working to challenge dominant configurations of social power (e.g. class, gender, race, or heteronormativity) and practicing feminist sensibility and reflexivity in their research (Pavlovskaya and St Martin, 2007). This research is particularly open to the qualitative potential of GIS because it aims to incorporate unprivileged and often non-measurable forms of experience not included in quantitative representations. Feminist GIS scholars have also worked to incorporate qualitative analytical functionality into GIS (see Knigge and Cope, 2006; also Knigge and Cope, this volume).

In this section, I suggest further possibilities for expanding the strengths of qualitative GIS. In particular, I consider how GIS can fruitfully enrich qualitative explanation by incorporating spatiality. I then discuss the recent work that exemplifies how qualitative GIS can visualize non-quantifiable experiences, feelings, and emotions; harness the rhetorical power of mapping by visualizing unprivileged ontologies; and ask questions that can only be answered through the combination of qualitative data and GIS-based analysis (that is, a 'mixed methods' approach).

Incorporating non-Cartesian spatiality

Concern with space and scale continue today in critical geography in the form of debates about the spatio-temporalities of human worlds (Harvey, 2006; Herod and Wright, 2002; Marston et al., 2005) as well as in GIScience in the form of ontologies research (Fonseca et al., 2000). The waves of 'spatial turn' have brought space as a key category into social sciences and humanities that also turn to using GIS and other geospatial technologies (Bol, 2004; Chambers et al., 2004; Pavlovskaya, forthcoming). This presents an ideal moment for GIS to realize its potential as a representational tool of critical geography. But a fundamental dilemma arises: GIS is associated with an absolute concept of space defined by a Cartesian grid, while critical human geography views space as produced by social relationships and experiences (see Harvey, 2006; Miller and Wentz, 2003). Can a GIS view space in anything but Cartesian terms?

Space is conventionally conceptualized in GIS as 'absolute', Euclidean, or Cartesian space that contains clearly defined objects with precise location and where processes operate on a number of fixed and analytically separate scales (e.g. local, regional, national, or global). This absolute space, associated with spatial science, enables formal analysis of spatial patterns and relationships, such as the distance decay function. As Miller and Wentz (2003) show, such conceptions prevail despite the fact that other representations of space within GIS are possible. Accordingly, GIS most often is used to do exactly this: to map and analyze spatial patterns in Euclidean space. Occasionally, it is used to visualize processes defined by relative positions of places and geographic objects, including connections, flows, networks, and movement. Mainstream GIS, however, has very limited capabilities for modeling flows and movement (mainly as cost–distance or network analysis). In critical geography, 'relational' space, along with time, is inseparable from social processes (Harvey, 2006; Massey, 1985) and may embrace non-measurable properties of place, human experience, and social power. Understanding these aspects of space requires qualitative modes of explanation prominent in Marxian, feminist, post-structural, and postcolonial approaches. GIS, however, is rarely used to represent 'relational space'. Furthermore, GIS does not represent people well because its objects are spatial features with attributes (e.g. discrete vector features or raster cells). It is difficult to model people's behavior or connect experiences to discrete spatial objects (Dorling, 1998; Kwan, 2004b; Miller, 2003; Openshaw, 1998; Poon, 2004). And people, obviously, are the main concern of human qualitative geography.

And yet, despite these major challenges, GIS also offers possibilities to qualitative modes of explanation. It does so precisely because it creates inherently spatial representations. It is possible, I believe, to find use for these representations in critical human geography or extend the representations themselves beyond the absolute space of spatial science. To do so, many important questions need to be addressed, within both human geography and GIS. How can we represent spatially complex connections, power relations, and collective meanings? How can the partiality of GIS representations open up to contestation and dialog with other partial representations? How can alternative mappings be created with, by, and for the disempowered social actors whose spatial experiences are not commonly represented? How can the authority of GIS-based representations be made less exclusive? How can the results of qualitative analysis of space be represented? How can we create powerful geographies of relational spaces using the absolute space of current GIS? Graphics often aptly communicate concepts, but representing a theoretical argument spatially is rare. The examples below illustrate some of these challenges.

Visualizing non-quantifiable experiences

In order to overcome the bias of GIS databases towards numerical information, feminist and other critical human geographers have begun to use unconventional spatial data such as narratives, in-depth interviews, hand-drawn maps, graphics, photographs, videos, as well as voices and sounds (Dorling, 1998; Kwan, 2002a; Sheppard, 2001). Using these methods, they create analytical representations of people's experiences, movements, and even such hard-to-quantify phenomena as emotions or webs of daily economic practices. Looking to model daily movement through urban space, Mei-Po Kwan (2002a) revived Torsten Hägerstrand's space–time geography approach which she applied to her analysis of women's daily travel. To implement GIS-based modeling of their movement, she combined urban land use and street network data with qualitative information from travel diaries kept by the respondents. Kwan visualized the three-dimensional life paths representing the daily travel of women from different ethnic and socio-economic groups. She concluded that not only are the uses of urban space gendered (a fact obscured by conventional urban models) but the differences between women from different class and racial backgrounds are also profound (Kwan, 1999a; 1999b; 2002a). In another project, Kwan (2007) visualized in a GIS the variations in safety of urban space as perceived by a Muslim woman after 11 September 2001. In this project, Kwan used emotion as a main type of data, acquired through ethnographic research, to be modeled and mapped.

In a project that explicitly combines GIS with an ethnographic study of low-income urban households, Matthews et al. (2005) designed a database that summarizes in-depth interview information and links it to places that people talk about in their interviews. This work has augmented the presentation of ethnographic data and added context by displaying census and crime information for the neighborhoods where the respondents lived. Matthews et al.'s work advances the interdisciplinary framework of a 'geoethnography' that combines geospatial technologies with urban ethnography. In a non-urban context, Hong Jiang (2003) combined an ethnographic study of villagers in Inner Mongolia with a remote sensing analysis of landscape change. She found that these approaches complemented each other such that she could weave a more compelling and complex story of landscape change.

Kevin St Martin (2005; 2008; forthcoming; and with Hall-Arber, 2007; in press) integrated GIS with ethnographic research while studying the potential for community management in the fishing industry of the US Northeast. In this participatory research project, community researchers (primarily women who were themselves fishers, fishermen's wives, or local advocates of their fishing communities) interviewed fishers about their fishing histories, communities, and local environmental knowledge using GIS maps as referents. National Marine Fisheries Service vessel trip report data (geocoded data reporting fishing trip locations) were analyzed quantitatively using density mapping and percentage volume contour (PVC) calculations to delineate the territories of particular fishing communities. Project participants were asked to comment on the accuracy and meaning of the resulting maps relative to community. Questions included whether project participants saw these fishing grounds as sites of common histories, shared knowledge, cooperation, and community formation; and whether or not the maps depicting a shared space produced a sense of community where none had previously existed.

In another example, Marie Cieri (2003) examined the sense of place produced by queer tourist industry propaganda in comparison to that directly experienced by lesbian tourists. She juxtaposed commercial tourist maps and tourist guide narratives with the hand-drawn spaces and stories told by her respondents (Figure 2.1).[2] She found that the queer tourist industry conflates lesbian and gay male spaces and reduces both to spaces of consumption, in contrast to the spaces with multiple meaning lived by the lesbian women.

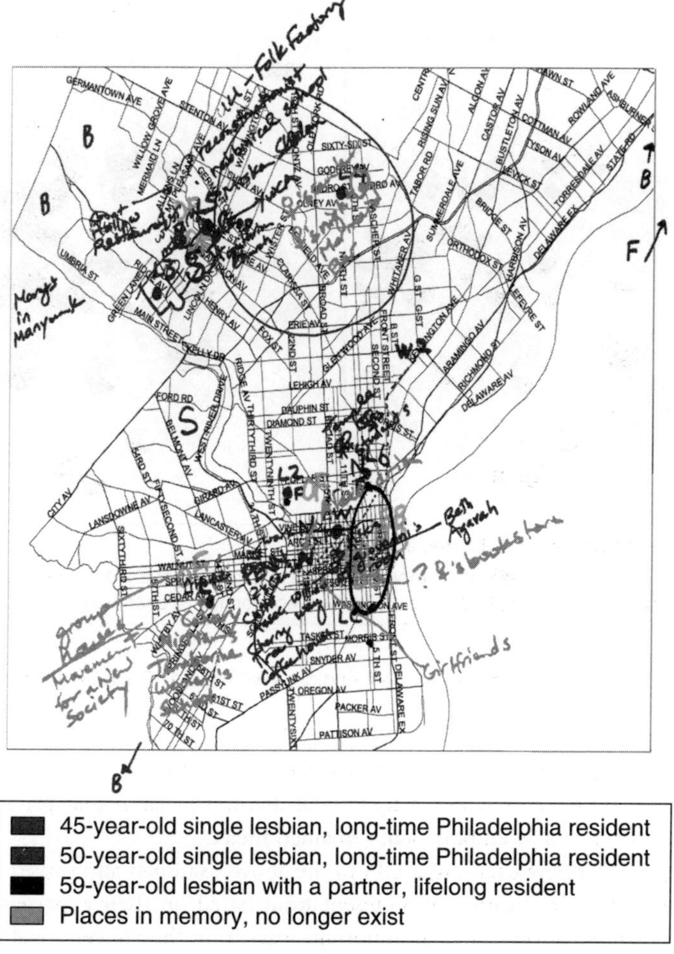

Figure 2.1 NW Philadelphia and center city: superimposed maps by three lesbians (Cieri, 2003)

In my own research on urban transformations under postsocialism (Pavlovskaya, 2002; 2004), I created maps of the multiple economies of Moscow households using ethnographic information from in-depth interviews (Figure 2.2). These maps show that in each household a wide range of economic activities is present both under socialism and especially, after the collapse of socialism, in the market-based economy. These activities included formal and informal employment for wages, informal and unpaid domestic production of goods and services (e.g. cooking, childcare), and exchanges of goods and services via networks of family and friends. While formal work for wages remains the primary concern of urban and economic policy and research, most other necessary and very time and labor intensive economic practices remain invisible and, therefore, under-theorized and ignored. Mapping networks of support in single- and two-parent households (Figure 2.3) also revealed that single parents were often successfully employed because they had to secure networks of

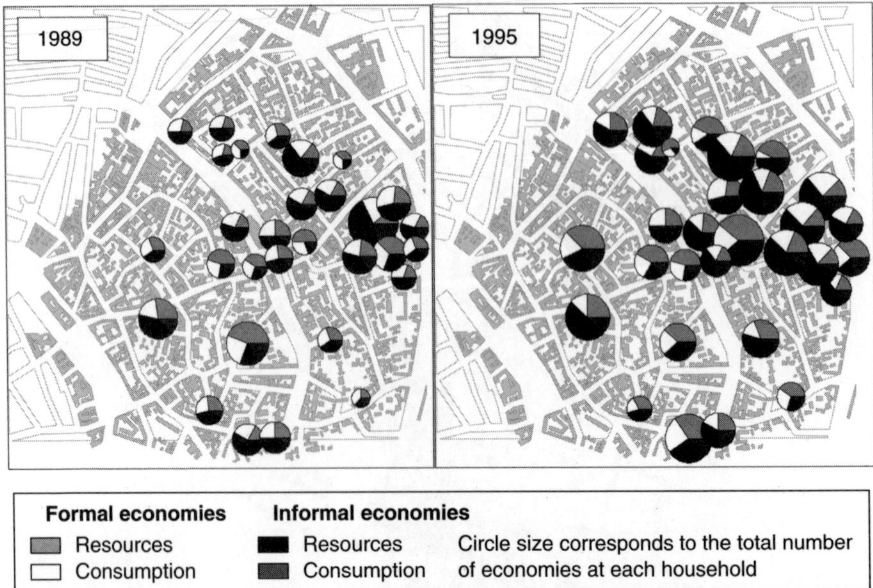

Figure 2.2 Multiple economies and households, 1989 and 1995, downtown Moscow

extended family and friends in order to have any kind of work. That was in contrast to two-parent households where a traditional division of labor that privileges male employment over female employment remained intact.

Integrating interview data into a GIS in the above examples also served to include the respondents as co-creators of representations based on their experiences. These alternative representations differ in important ways from the conventional depictions of economies, households, danger and crime, natural resources, social services, or consumption patterns that are based on indicators computed from large and impersonal statistical datasets.

Visualizing unprivileged ontologies

No less important is a visualization strategy that creates social ontologies that are invisible for conventional theories and methods. Mapping such phenomena, relationships, and landscapes (e.g. the daily paths of women, experiences of Muslim women, territories at sea used communally, lived lesbian spaces, informal household economies, or daily networks of support) makes them visible and, therefore, 'real' and significant theoretically and politically. In other words, 'positioning' these unprivileged phenomena on the map using a GIS that merges scientific authority with visual impact performs an ontological function: it 'creates' the landscapes produced by these processes and legitimizes them. The power of GIS to constitute such worlds is particularly appealing for critical geography because of its concern with including and representing the excluded.

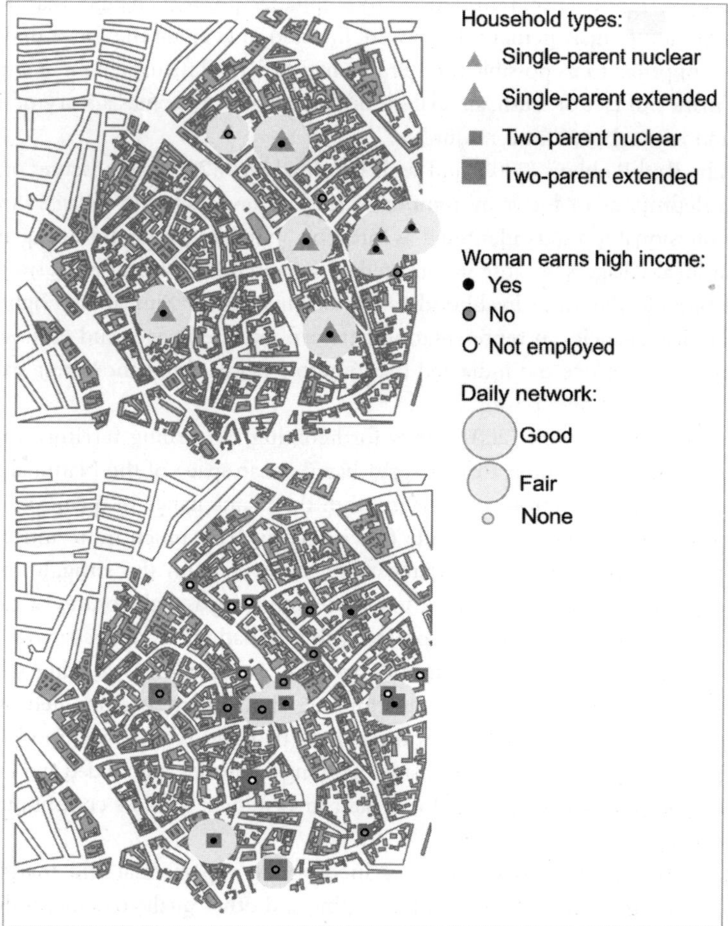

Figure 2.3 Household support networks, 1995, downtown Moscow

Mixed methods

Thinking of a qualitative GIS as a mixed method opens further possibilities. The effect of combining quantitative and qualitative methods with geospatial technologies goes beyond gaining more by adding different types of knowledge or even complementing partial knowledges. Mixing these methods can achieve two more important (although related) goals. One is the ability to ask research questions that could not be asked if only one method is used. The second is to actually look for inconsistencies in partial knowledges produced by different techniques and treat them as research opportunities, as opposed to error or incompleteness of data. In this case, discrepancies become openings into an inquiry about social power configurations that produce these different representations and their effects.

Feminist political ecologist Andrea Nightingale (2003) specifically focused on the inconsistencies in the accounts of changes in forest cover based on aerial photography and the ecological histories of villagers in Nepal. She found that villagers participating in community forestry programs tended to emphasize positive changes that occurred under community management. They were invested in keeping the forest under local control as opposed to its possible transfer to a national-level management. Rather than being a matter of fact or truth, an analysis of discrepancies becomes a story of political power and control over local resources.

Work by Paul Robbins (2003) and with Tara Maddock (2000) also focuses on differences in definitions of forest by remote sensing professionals and villagers in India. What professional foresters identified as forest on a satellite image did not qualify as a forest for local villagers because it consisted of replanted (indeed, invasive) species that did not provide the same livelihood as the original forest. Similar to Nightingale's work, the discrepancies in representation between satellite images and ethnographies of community resource use indicated that multiple truths about the 'forest' expressed a politics of control over resources.

Kevin St Martin's work (2001; 2008; forthcoming) on fishing territories in New England reveals, for example, that the grid-based ocean space of the National Marine Fisheries Service comes from its concern with the maintenance of quantities of fishing stock in a borderless ocean, while fishers' oceans have much more intricate and complicated geographies. These discrepancies are evident in the struggles between fishing communities and government management over seasonal closures of particular areas. The closures are designed to protect fish populations from predacious fishers, who are thought to be endlessly mobile individuals capable of catching unlimited quantities of fish. This is in contrast to thinking about them as embodied men and women who fish in particular places they know best and whose livelihood depends on access to these places. This second vision of fishing territories as bounded and harvested by local communities makes a case for greater involvement of these communities into fisheries management.

These examples show that GIS may incorporate experiential and marginalized spatialities that are best elicited by ethnographic and other qualitative methods. Using mixed methods, GIS also opens the inconsistencies in data derived from different sources to investigation of the social power dynamics that produce different representations. In other words, GIS may provide ways to address relational spaces of power, whether they are or are not bound to a Cartesian grid. These questions are at the core of current human geography concerns.

CONCLUSION

This chapter approached the subject of using GIS in qualitative research by treating GIS, similar to other research methods, as a power relation. The dominant view of GIS as a quantitative technology, then, is not grounded in its innate properties but is a result of negotiations between differing practices of knowledge production. The critical examination of the functionality of GIS presented in this chapter reveals that in many ways GIS is intimate with non-quantitative data and modes of analysis, while its application in quantitative geography and spatial analysis, has been surprisingly limited. Most academic and other users rely on its areas of functionality that can serve qualitative researchers equally well, such as visualization, integration of different types of data, querying, and basic spatial analysis.

The challenge is to open GIS to qualitative research so that complex relationships, non-quantifiable properties, unprivileged ontologies, and fluid human worlds can be represented spatially and better understood. Re-imagining GIS as a flexible tool for creating diverse human geographies not solely confined to the 'absolute space' of spatial science has already begun. As the examples above show, GIS could be used by critical human geographers engaged in qualitative research and focusing on relational spaces of social power. While it is far from providing answers to all questions, GIS can be fruitfully combined with other research strategies. It can incorporate experiences elicited through ethnographic work and other qualitative research methods. It can use non-quantitative data (such as images, video, sound, narrative) in combination with more standard datasets, such as census data (for an example, see Jung, this volume). As a powerful representational tool, GIS can reconstitute unprivileged social ontologies by placing them within the authoritative field of science and technology. It also enables mixed methods approaches that integrate geospatial technologies with qualitative and quantitative research. And, finally, as a mixed methods medium GIS encourages researchers to seek to understand power dynamics and authority clashes that produce always partial and often conflicting spatial representations of human worlds.

ACKNOWLEDGEMENTS

I am grateful to the editors, Sarah Elwood and Meghan Cope, as well as the two anonymous reviewers and Kevin St Martin for their constructive and very helpful comments. Thinking and writing time for this chapter was made possible in part by funding from the Gender Equity Project at Hunter College (NSF ADVANCE grant).

NOTES

1 Parts of this chapter, especially the section on 'openings', were previously published by Pion Limited, London in Pavlovskaya, M.E. (2006) 'Theorizing with GIS: a tool for critical geographies?', *Environment and Planning* A, 38 (11): 2003–20. Here they appear thoroughly revised and with the addition of new content.

2 All figures reproduced with permission of the authors.

REFERENCES

Ahlqvist, O. (2004) 'A parameterized representation of uncertain conceptual spaces', *Transactions in GIS*, 8 (4): 493–514.

Amin, A. and Thrift, N. (2000) 'What kind of economic theory for what kind of economic geography?', *Antipode*, 32: 4–9.

Armstrong, M.P. and Ruggles, A.J. (2005) 'Geographic information technologies and personal privacy', *Cartographica*, 40 (4): 63–73.

Babbie, E.R. (2000) *The Practice of Social Research,* 9th edn. Belmont, CA: Wadsworth.

Barnes, T.J. (1998) 'A history of regression: actors, networks, numbers and machines', *Environment and Planning A*, 30 (2): 203–23.

Barnes, T.J. (2000) 'Inventing Anglo-American economic geography, 1889–1960', in E.S. Sheppard and T.J. Barnes (eds), *A Companion to Economic Geography*. London: Blackwell. pp. 11–26.

Barnes, T.J. (2001) 'Lives lived and lives told: biographies of geography's quantitative revolution', *Environment and Planning D: Society and Space*, 19 (4): 409–29.

Baxter, J. and Eyles, J. (1997) 'Evaluating qualitative research in social geography: establishing "rigour" in interview analysis', *Transactions of the Institute of British Geographers*, 22 (4): 505–25.

Bell, S. and Reed, M. (2004) 'Adapting to the machine: integrating GIS into qualitative research', *Cartographica,* 39 (1): 55–66.

Bol, P. (2004) 'Putting history in geography: the challenges of creating a GIS for the history of China', paper presented at History and Geography: Assessing the Role of Geographic Information in Historical Scholarship, Chicago, IL.

Chambers, K.J., Corbett, J., Keller, C.P. and Wood, C.J.B. (2004) 'Indigenous knowledge, mapping and GIS: a diffusion of innovation perspective', *Cartographica*, 39 (3): 19–31.

Cieri, M. (2003) 'Between being and looking: queer tourism promotion and lesbian social space in Greater Philadelphia', *ACME: An International E-Journal for Critical Geographies*, 2 (2): 147–66.

Clarke, K.C. (1999) *Getting Started with Geographic Information Systems*, 2nd edn. Upper Saddle River, NJ: Prentice-Hall.

Clarke, K.C., Parks, B.O. and Crane, M.P. (2002) *Geographic Information Systems and Environmental Modeling*. Upper Saddle River, NJ: Prentice-Hall.

Clifford, N.J. and Valentine, G. (eds) (2003) *Key Methods in Geography*. London: Sage.

Cloke, P., Philo, C. and Sadler, D. (1991) *Approaching Human Geography: An Introduction to Contemporary Theoretical Debates*. New York: Guilford.

Cloke, P., Crang, P., Goodwin, M., Painter, J. and Philo, C. (2004) *Practising Human Geography*. London: Sage.

Cosgrove, D.E. and Daniels, S. (eds) (1988) *The Iconography of Landscape: Essays on the Symbolic Representation, Design, and Use of Past Environments*. Cambridge: Cambridge University Press.

Craig, W.J., Harris, T.M. and Weiner, D. (eds) (2002) *Community Participation and Geographic Information Systems*. London: Taylor and Francis.

Crampton, J.W. (2001) 'Maps as social constructions: power, communication, and visualization', *Progress in Human Geography*, 25 (2): 235–52.

Crampton, J.W. (2003) 'Cartographic rationality and the politics of geosurveillance and security', *Cartography and Geographic Information Science*, 30 (2): 135–48.

Crang, M. (2002) 'Qualitative methods: the new orthodoxy? Progress report', *Progress in Human Geography*, 26 (5): 647–55.

Creswell, J.W. (2003) *Research Design: Qualitative, Quantitative, and Mixed Methods Approaches*, 2nd edn. Thousand Oaks, CA: Sage.

Crisman, N. (2002) *Exploring Geographic Information Systems*, 2nd edn. New York: Wiley.

Curry, M.R. (1997) 'The digital individual and the private realm', *Annals of the Association of American Geographers*, 87 (4): 681–99.

De Smith, M.J., Goodchild, M.F. and Longley, P.A. (2007) *Geospatial Analysis: A Comprehensive Guide to Principles, Techniques and Software Tools*. Leicester: Troubador.

Dixon, D.P. and Jones III, J.P. (1996) 'For a supercalifragilisticexpialodocious scientific geography', *The Annals of the Association of American Geographers*, 86 (4): 767–79.

Dobson, J. E. and Fisher, P. F. (2003) 'Geoslavery', *IEEE Technology and Society Magazine*, Spring: 47–52.

Doel, M.A. (2001) 'Qualified quantitative geography', *Environment and Planning D: Society and Space*, 19 (5): 555–72.

Dorling, D. (1998) 'Human cartography: when it is good to map', *Environment and Planning A*, 30 (2): 277–88.

Eastman, J.R. (2003) *IDRISI Kilimanjaro Guide to GIS and Image Processing*. Worcester, MA: IDRISI production, Clark University.

Edney, M. (1997) *Mapping the Empire: The Geographical Construction of British India*. Chicago: University of Chicago Press.

Elwood, S.A. (2006a) 'Negotiating knowledge production: the everyday inclusions, exclusions, and contradictions of participatory GIS research', *The Professional Geographer*, 58 (2): 197–208.

Elwood, S.A. (2006b) 'Beyond cooptation or resistance: urban spatial politics, community organizations, and GIS-based spatial narratives', *Annals of the Association of American Geographers*, 92 (6): 323–41.

Elwood, S.A. and Leitner, H. (1998) 'GIS and community based planning: exploring the diversity of neighborhood perspectives and needs', *Cartography and Geographic Information Systems*, 25: 77–88.

Elwood, S.A. and Martin, D.G. (2000) '"Placing" interviews: location and scales of power in qualitative research', *The Professional Geographer*, 52 (4): 649–57.

Flowerdew, R. (1998) 'Reacting to *Ground Truth*', *Environment and Planning A*, 30 (2): 289–301.

Fonseca, F., Egenhofer, M., Davis, C. and Borges, K. (2000) 'Ontologies and knowledge sharing in urban GIS', *Computer, Environment and Urban Systems*, 24 (3): 232–51.

Fotheringham, A.S. (1997) 'Trends in quantitative methods I: stressing the local', *Progress in Human Geography*, 21 (1): 88–96.

Fotheringham, A.S., Brunsdon, C. and Charlton, M. (2002) *Geographically Weighted Regression: The Analysis of Spatially Varying Relationships*. Chichester: Wiley.

Foucault, M. (1980) *Power/Knowledge: Selected Interviews and Other Writings, 1972–1977*. New York: Pantheon.

Getis, A. and Ord, J.K. (1996) 'Local spatial statistics: an overview', in P. Longley and M. Batty (eds), *Spatial Analysis: Modelling in a GIS Environment*. Cambridge: Geoinformation International.

Gibson-Graham, J.K. (2000) 'Poststructuralist interventions', in E.S. Sheppard and T.J. Barnes (eds), *A Companion to Economic Geography*. London: Blackwell. pp. 95–110.

Gilbert, M.R. and Masucci, M. (2004) 'Moving beyond "gender and GIS" to a feminist perspective on information technologies: the impact of welfare reform on women's IT needs', in J. Seager and L. Nelson (eds), *A Companion to Feminist Geography*. London: Blackwell. pp. 305–21.

Goodchild, M.F. (1991) 'Just the facts', *Political Geography Quarterly*, 10: 335–7.

Goss, J. (1995) 'Marketing the new marketing: the strategic discourse of geodemographic information systems', in J. Pickles (ed.), *Ground Truth: The Social Implications of Geographic Information Systems*. New York: Guilford. pp. 130–70.

Graves, S.M. (2003) 'Landscapes of predation, landscapes of neglect: a locational analysis of payday lenders and banks', *The Professional Geographer*, 55 (3): 303–17.

Hanson, S. (1997) 'As the world turns: new horizons in feminist geographic methodologies', in J.P. Jones III, H.J. Nast and S.M. Roberts (eds), *Thresholds in Feminist Geography: Difference, Methodology, Representation*. Lanham, MD: Rowman and Littlefield. pp. 119–28.

Hanson, S. (2002) 'Connections', *Gender, Place and Culture: A Journal of Feminist Geography*, 9 (3): 301–03.

Hanson, S., Kominiak, T. and Carlin, S. (1997) 'Assessing the impact of location on women's labor market outcome: a methodological exploration', *Geographical Analysis*, 2 (4): 282–97.

Haraway, D. (1991) 'Situated knowledges: the science question in feminism and the privilege of partial perspective', in *Simians, Cyborgs and Women: The Reinvention of Nature*. London: Free Association Books. pp. 183–201.

Harley, J.B. (1992) 'Deconstructing the map', in T.J. Barnes and J.S. Duncan (eds), *Writing Worlds: Discourse, Text, and Metaphor in the Representation of Landscape*. New York: Routledge. pp. 231–47.

Harvey, D. (2006) 'Space as a Key Word', in N. Castree and D. Gregory (eds), *David Harvey: A Critical Reader*. London: Blackwell. pp. 270–94.

Herod, A. and Wright, M.W. (eds) (2002) *Geographies of Power: Placing Scale*. London: Blackwell.

Jiang, H. (2003) 'Stories remote sensing images can tell: integrating remote sensing analysis with ethnographic research in the study of cultural landscapes', *Human Ecology: An Interdisciplinary Journal*, 31 (2): 215–32.

Jones III, J.P. and Dixon, D.P. (1998) 'My dinner with Derrida, or spatial analysis and post-structuralism do lunch', *Environment and Planning A*, 30: 247–60.

Katz, C. (1992) 'All the world is staged: intellectuals and the projects of ethnography', *Environment and Planning D: Society and Space*, 10: 495–510.

Katz, C. (2001) 'On the grounds of globalization: a topography for feminist political engagement', *Signs: Journal of Women in Culture and Society*, 26 (4): 1213–35.

Knigge, L. and Cope, M. (2006) 'Grounded visualization: integrating the analysis of qualitative and quantitative data through grounded theory and visualization', *Environment and Planning A*, 38 (11): 2021–37.

Kwan, M.P. (1999a) 'Gender and individual access to urban opportunities: a study using space–time measures', *The Professional Geographer*, 51 (2): 210–27.

Kwan, M.P. (1999b) 'Gender, the home–work link, and space–time patterns of non-employment activities', *Economic Geography*, 75: 370–94.

Kwan, M.P. (2002a) 'Feminist visualization: re-envisioning GIS as a method in feminist geographic research', *Annals of the Association of American Geographers*, 92 (4): 645–61.

Kwan, M.P. (2002b) 'Is GIS for women? Reflections on the critical discourse in the 1990s', *Gender, Place and Culture: A Journal of Feminist Geography*, 9 (3): 271–9.

Kwan, M.P. (2002c) 'Is GIS for women? A critical reflection', paper presented at the Department of Geography, Rutgers University, New Brunswick.

Kwan, M.P. (2004a) 'Beyond difference: from canonical geography to hybrid geographies', *Annals of the Association of American Geographers*, 94 (4): 756–63.

Kwan, M.P. (2004b) 'GIS methods in time-geographic research: geocomputation and geovisualization of human activity patterns', *Geografiska Annaler, Series B: Human Geography*, 86 (4): 267–80.

Kwan, M.P. (2007) 'Affecting geospatial technologies: toward a feminist politics of emotion', *The Professional Geographer*, 59 (1): 22–34.

Kwan, M.P. and Knigge, L. (2006) 'Doing qualitative research using GIS: an oxymoronic endeavor? Editorial', *Environment and Planning A*, 38 (11): 1999–2002.

Lake, R.W. (1993) 'Planning and applied geography: positivism, ethics, and geographic information systems', *Progress in Human Geography*, 17: 404–13.

Lawson, V. (1995) 'The politics of difference: examining the quantitative/qualitative dualism in post-structuralist feminist research', *The Professional Geographer*, 47 (4): 449–57.

Lewis, G.M. (ed.) (1998) *Cartographic Encounters: Perspectives on Native American Mapmaking and Map Use*. Chicago: University of Chicago Press.

Livingstone, D.N. (1992) *The Geographical Tradition: Episodes in the History of a Contested Enterprise*. Oxford: Blackwell.

Longley, P.A., Goodchild, M.F., Maguire, D.J. and Rhind, D.W. (2005) *Geographic Information Systems and Science*, 2nd edn. New York: Wiley.

MacEachren, A. (1994) 'Visualization in modern cartography: setting the agenda', in A. MacEachren and D. Taylor (eds), *Visualization in Modern Cartography*. Oxford: Pergamon. pp. 1–12.

MacEachren, A.M., Wachowicz, M., Edsall, R.M., Haug, D. and Masters, R. (1999) 'Constructing knowledge from multivariate spatiotemporal data: integrating geographical visualization with knowledge discovery in database methods', *International Journal of Geographical Information Science*, 13 (4): 311–34.

MacEachren, A.M., Gahegan, M. and Pike, W. (2004) 'Visualization for constructing and sharing geo-scientific concepts', *Proceedings of the National Academy of Sciences of the United States of America*, 101: 5279–86.

Margai, F.L. (2001) 'Health risks and environmental inequality: a geographical analysis of accidental releases of hazardous materials', *The Professional Geographer*, 53 (3): 422–34.

Marston, S.A., Jones III, J.P. and Woodward, K. (2005) 'Human geography without scale', *Transactions of the Institute of British Geographers*, 30: 416–32.

Massey, D. (1985) 'New directions in space', in D. Gregory and J. Urry (eds), *Social Relations and Spatial Structures*. New York: St Martin's. pp. 9–19.

Matthews, S., Detwiler, J. and Burton, L. (2005) 'Geo-ethnography: coupling geographic information analysis techniques with ethnographic methods in urban research', *Cartographica*, 40 (4): 75–90.

McLafferty, S.L. (1995) 'Counting for women', *The Professional Geographer*, 47 (4): 436–42.

McLafferty, S.L. (2002) 'Mapping women's worlds: knowledge, power, and the bounds of GIS', *Gender, Place and Culture: A Journal of Feminist Geography*, 9 (3): 263–9.

McLafferty, S.L. (2005) 'Women and GIS: geospatial technologies and feminist geographie', *Cartographica*, 40 (4): 37–45.

Miller, H. (2003) 'What about people in geographic information science?', *Computers, Environment and Urban Systems*, 27 (5): 447–53.

Miller, H. and Wentz, E. (2003) 'Representation and spatial analysis in geographic information systems', *Annals of the Association of American Geographers*, 93 (3): 574–94.

Nightingale, A. (2003) 'A feminist in the forest: situated knowledges and mixing methods in natural resource management', *ACME: An International E-Journal for Critical Geographies*, 2 (1): 77–90.

Openshaw, S. (1991) 'A view on the GIS crisis in geography, or using GIS to put Humpty Dumpty back again', *Environment and Planning A*, 23: 621–8.

Openshaw, S. (1998) 'Towards a more computationally minded scientific human geography', *Environment and Planning A*, 30: 317–32.

Openshaw, S. and Taylor, P.J. (1979) 'A million or so correlated coefficients: three experiments on the modifiable areal unit problem', in N. Wrigley and R.J. Bennet (eds), *Statistical Applications in the Spatial Sciences*. London: Pion.

Pavlovskaya, M.E. (2002) 'Mapping urban change and changing GIS: other views of economic restructuring', *Gender, Place and Culture: A Journal of Feminist Geography*, 9 (3): 281–9.

Pavlovskaya, M.E. (2004) 'Other transitions: multiple economies of Moscow households in the 1990s', *Annals of the Association of American Geographers*, 94 (2): 329–51.

Pavlovskaya, M.E. (2006) 'Theorizing with GIS: a tool for critical geographies?', *Environment and Planning A*, 38 (11): 2003–20.

Pavlovskaya, M.E. (forthcoming) 'Feminism, maps, and GIS', in R. Kitchin and N. Thrift (eds), *International Encyclopedia of Human Geography*. Elsevier.

Pavlovskaya, M.E. and St Martin, K. (2007) 'Feminism and GIS: from a missing object to a mapping subject', *Geography Compass*, 1 (3): 583–606.

Peng, Z.R. (2001) 'Internet GIS for public participation', *Environment and Planning B: Planning and Design*, 28 (6): 889–906.

Pickles, J. (ed.) (1995) *Ground Truth: The Social Implications of Geographic Information Systems*. New York: Guilford.

Pickles, J. (1997) 'Tool or science? GIS, technoscience, and the theoretical turn', *Annals of the Association of American Geographers*, 87 (2): 363–72.

Pickles, J. (2004) *A History of Spaces: Cartographic Reason, Mapping, and the Geo-Coded World*. London: Routledge.

Plummer, P.S. (2000) 'The modeling tradition', in E.S. Sheppard and T.J. Barnes (eds), *A Companion to Economic Geography*. London: Blackwell. pp. 27–40.

Plummer, P.S. and Sheppard, E.S. (2001) 'Must emancipatory economic geography be qualitative?', *Antipode*, 33 (2): 194–200.

Poon, J.P.H. (2004) 'Quantitative methods: past and present', *Progress in Human Geography*, 28 (6): 807–14.

Robbins, P. (2003) 'Beyond ground truth: GIS and the environmental knowledge of herders, professional foresters, and other traditional communities', *Human Ecology*, 31 (2): 233–53.

Robbins, P. and Maddock, T. (2000) 'Interrogating land cover categories: metaphor and method in remote sensing', *Cartography and Geographic Information Science*, 27 (4): 295–309.

Robinson, A.H. and Petchenik, B.B. (1976) *The Nature of Maps*. Chicago: University of Chicago Press.

Rocheleau, D. (1995) 'Maps, numbers, text, and context: mixing methods in feminist political ecology', *The Professional Geographer*, 47 (4): 458–66.

Rose, G. (1992) 'Geography as a science of observation: the landscape, the gaze, and masculinity', in F. Driver and G. Rose (eds), *Nature and Science: Essays in the History of Geographical Knowledge*. London: Historical Geography Research Group. pp. 8–18.

Schroeder, P. (1996) 'Criteria for the design of a GIS/2', http://ncgia.spatial.maine.edu/ppgis/criteria.html, accessed 2 March 2004.

Schuurman, N. (1999) 'Speaking with the enemy? Guest editorial and interview with Michael Goodchild', *Environment and Planning D: Society and Space*, 17 (1): 1–15.

Schuurman, N. (2000) 'Trouble in the heartland: GIS and its critics in the 1990s', *Progress in Human Geography*, 24 (4): 569–90.

Schuurman, N. (2002) 'Women and technology in geography: a cyborg manifesto for GIS', *The Canadian Geographer*, 46 (3): 262–5.

Schuurman, N. and Pratt, G. (2002) 'Care of the subject: feminism and critiques of GIS', *Gender, Place and Culture: A Journal of Feminist Geography*, 9 (3): 291–9.

Scott, M., Cutter, S.L., Menzel, C. and Ji, M. (1997) 'Spatial accuracy of the EPA's environmental hazards databases and their use in environmental equity analysis', *Applied Geographic Studies*, 1 (1): 45–61.

Sheppard, E.S. (1993) 'Automated geography: what kind of geography for what kind of society?', *The Professional Geographer*, 45: 457–60.

Sheppard, E.S. (2001) 'Quantitative geography: representations, practices, and possibilities', *Environment and Planning D: Society and Space*, 19 (5): 535–54.

Sheppard, E.S. (2005) 'Knowledge production through critical GIS: genealogy and prospects', *Cartographica*, 40 (4): 5–21.

Sieber, R.E. (2004) 'Rewiring for GIS/2', *Cartographica*, 39 (1): 25–39.

Smith, N. (1992) 'History and philosophy of geography: real wars, theory wars', *Progress in Human Geography*, 16: 257–71.

Sparke, M. (1998) 'A map that roared and the original atlas: Canada, cartography, and the narration of nation', *Annals of the Association of American Geographers*, 88 (3): 463–95.

Staeheli, L.A. and Mitchell, D. (2005) 'The complex politics of relevance in geography', *Annals of the Association of American Geographers*, 95 (2): 357–72.

St Martin, K. (1995) 'Changing borders, changing cartography: Possibilities for intervening in the new world order', in A. Callari, S. Cullenberg and C. Biewener (eds), *Marxism in the Postmodern age: Confronting the New World Order*. New York: Guilford. pp. 459–68.

St Martin, K. (2001) 'Making space for community resource management in fisheries', *Annals of the Association of American Geographers*, 91 (1): 122–42.

St Martin, K. (2005) 'Mapping economic diversity in the first world: the case of fisheries', *Environment and Planning A*, 37: 959–79.

St Martin, K. (2008) 'Mapping community use of fisheries resources in the U.S. Northeast', *Journal of Maps*, 38–41.

St Martin, K. (forthcoming) 'Quantitative and critical GIS methods to foster community participation in natural resource management', *The Professional Geographer*.

St Martin, K. and Hall-Arber, M. (2007) 'Environment and development: (re)connecting community and commons in New England fisheries', in S. Kindon, R. Pain and M. Kasby (eds), *Connecting People, Participation and Place: Participatory Action Research Approaches and Methods*. New York: Routledge. pp. 51–9.

St Martin, K. and Hall-Arber, M. (in press) 'The missing layer: geotechnologies, communities, and implications for marine spatial planning', *Marine Policy*.

St Martin, K. and Pavlovskaya, M. (forthcoming a) 'Secondary data: engaging numbers critically', in J.P. Jones III and B. Gomez (eds), *Research Methods in Geography: A First Course*. London: Blackwell.

St Martin, K. and Pavlovskaya, M. (forthcoming b) 'Ethnography', in N. Castree, D. Demeritt, D. Liverman and B. Rhoads (eds), *Companion to Environmental Geography*. London: Blackwell.

St Martin, K. and Wing, J. (2007) 'The discourse and discipline of GIS', *Cartographica*, 42 (3): 235–48.

Strauss, A.L. (1995) *Qualitative Analysis for Social Scientists*. Cambridge, MA: Cambridge University Press.

Sui, D.Z. (2000) 'Visuality, aurality, and shifting metaphors of the geographical thought in the late twentieth century', *Annals of the Association of American Geographers*, 90 (2): 322–43.

Sui, D.Z. and Morrill, R. (2004) 'Computers and geography: from automated geography to digital Earth', in S.D. Brunn, S.L. Cutter and J.W. Harrington Jr (eds), *Geography and Technology*. Berlin: Springer. pp. 81–108.

Tashakkori, A. and Teddlie, C. (eds) (2002) *Handbook of Mixed Methods in Social and Behavioral Research*. Thousand Oaks, CA: Sage.

Treves, V. (2005) 'Towards a law enforcement technologies complex: situating Compstat in neo-liberal penality'. Master's thesis, Hunter College, New York.

Wang, F. (2000) 'Modeling commuting patterns in Chicago in a GIS environment: a job accessibility perspective', *The Professional Geographer*, 52 (1): 120–33.

Wong, D.W.S. (2003) 'Spatial decomposition of segregation indices: a framework toward measuring segregation at multiple levels', *Geographical Analysis*, 35 (3): 179–95.

Wong, D. W. S. (2004) 'The modifiable areal unit problem (MAUP)', in D.G. Janelle, B. Warf and K. Hansen (eds), *Worldminds: Geographical Perspectives on 100 Problems*. Dordrecht: Kluwer. pp. 571–8.

Wright, D. J., Goodchild, M.F. and Proctor, J.D. (1997) 'Demystifying the persistent ambiguity of GIS as "tool" versus "science"?', *Annals of the Association of American Geographers*, 87 (2): 346–62.

REPRESENTATIONS

3

METADATA AS A SITE FOR IMBUING GIS WITH QUALITATIVE INFORMATION

Nadine Schuurman

ABSTRACT

Whilst there is consensus that GISystems should be able to integrate and analyze qualitative data, there are few mechanisms for achieving this goal. Metadata (data about data) constitute a potentially rich repository of qualitative information about both spatial and non-spatial attributes in databases. Metadata have traditionally been limited to few fields, focused exclusively on geometric attributes, and been largely ignored by data users. Low utilization of metadata is attributable to low completion rates of metadata forms by data producers as well as the limited nature of the metadata provided. This chapter illustrates the potential for metadata to become a repository of qualitative information about both quantitative and qualitative spatial and non-spatial attributes. In the process, metadata will enhance the ability of GIS users to incorporate multiple knowledge systems – or ontologies. Metadata are a means of identifying a distinct ontology associated with a given dataset. A method for eliciting *ontology-based metadata* from data stewards based on ethnography is reviewed. Eight fields that capture qualitative information about data are described. These form an infrastructure or platform for the introduction of qualitative metadata in a computational environment. Two examples of the population of an ontology-based metadata table with data from health registries and forest cover records are given as illustrations of how qualitative metadata can enrich GIS analysis.

INTRODUCTION

Researchers remain hopeful that all spatial issues can be richly represented, whether they concern the environmental impact of the development of a forested area or the measurement of the health of newborn babies across a large geographical area. At present, however, there are impediments to achieving such a dream. Integration of rich datasets from multiple perspectives is key to a vision of a fair and equitable GIS. One goal of being able to use multiple datasets is to include many perspectives and different knowledge systems within a singular GIS. Each of these knowledge systems represents different ways of looking at the world with attendant agendas. Each of these ways of knowing, understanding, and organizing the world can be thought of as an *ontology*.

At present, one of the chief barriers to representing multiple ways of looking at the world – or different ontologies – is the data we work with. Data – as they appear in databases – are usually sparse, lacking contextual information. Data collection and use are based on social practices, yet once data are encoded in tables, their social context is forgotten. While individual researchers may be aware of the contextual background to a dataset, most users treat data as if they were the truth about the world. This results in problems with data sharing across institutions as well as data interpretation by GIS users. To date, there has been no concerted effort to link context to data – in a formal manner. Rather, data sit in isolated boxes in databases with semantic interpretation left largely in the hands of users. Of course, many data collectors have carefully explained the limitations and strengths of data to new users; a formal structure for this qualitative background to data is the goal of this chapter and *ontology-based metadata*.

The chapter begins by explaining what ontologies are and how they are related to epistemology, goes on to describe what metadata are, how metadata can be used to distinguish ontologies, and how ontology-based metadata can be collected, and ends by presenting two examples of how metadata can enhance the understanding of different datasets that describe the same geographical space and issues, but in different ways.

ONTOLOGIES: ORGANIZATION OF THE WORLD IN A COMPUTABLE UNIVERSE

Ontology as a concept is very confusing as it means different things to human geographers and social theorists than it does to information scientists. Ontologies in a GIS context are quite different from those in a philosophical context. Ontology in a traditional philosophical interpretation refers to the 'essence of being', an elemental reality that defines a phenomenon (Gregory, 1994; 2000b). There is an assumption that ontology is singular in this philosophical tradition. In computing and information sciences, however, the meaning of the word 'ontology' has been adapted in order to describe data and programming realms. In an information science context, ontology refers to a fixed universe of discourse in which all entities (e.g. spatial phenomena such as mountains, rivers, towns, and buildings) are defined and the possible relationships between each entity are also prescribed (Gruber, 1995; Smith and Mark, 2001). Each ontology in this context refers to a way of seeing the world or, more practically, as a classification system. If you think of a geological map, then everything that is on the map legend (e.g. sediments, metamorphic layers, and igneous rock) constitutes the ontology of the database that was used to create the map.

In the computing context, we can have multiple ontologies (e.g. multiple legends and classification systems). Indeed every new dataset or classification system constitutes an ontology as each one describes different spatial entities with different explicit relationships or interactions (Agarwal, 2005). Interactions between spatial entities associated with a particular ontology are proscribed in a computational environment (Bittner and Edwards, 2001; Brodaric and Gahegan, 2001; Fonseca et al., 2002; Frank, 2001; Winter, 2001). So, for instance, roads may not cross rivers in the absence of a bridge.

Spatial objects represented in a GIS require digital encoding of the entities and also encoding of the relationships between them. For example, a road can cross a bridge, but a river must run under it, and bus stops occur only along bus routes. Likewise, environmental modeling might restrict bears to watersheds that travel along eco-corridors (ideally) while humans traverse roads. These rules of interaction between entities are intrinsic to their definition and comprise an important part of the ontology.

Epistemology is closely related to ontology. How one studies and understands the world contributes to the ontologies that are evident to the observer and become encoded in a dataset. The epistemology used to develop or interpret an ontology (collection of things with clearly defined interrelationships) has a profound effect on its interpretation. Realism and positivism are two well-known philosophical branches of epistemology (Gregory, 2000a; 2000b). Realism is a way of looking at the world that assumes that there are invisible structures that pattern the course of events. For instance, social policies that encourage resource depletion or increases to social assistance payments uptake are examples of structures that precipitate events (e.g. clear cuts or fewer homeless people) (Schuurman, 2004). Positivism, on the other hand, is based on the premise that theories can be developed from observations or empiricism. Both branches of epistemology, however, can be hijacked by underlying agendas. Therefore a more pragmatic way of viewing epistemology is as a lens through which the observer or researcher organizes events and entities on the earth's surface.

A set of spatial entities such as forest, lake, river and meadow can be viewed from more than one epistemological perspective. A logging company might regard this set of geographical features as potential resources while an environmentalist might perceive the same landscape as a habitat for animals or an ecological system. Each group will construct their resulting computational ontologies differently based on different epistemologies. Epistemology imbues geographical objects with different meanings with different resultant ontologies, in the information science sense of the world.

We can think of ontologies as multiple, self-contained, logical systems – each with its own epistemological base. Each of these computable ontologies is socially constructed, just like the data they are populated with. For example, there are several government and non-profit agencies in the Canadian province of British Columbia that describe forest cover data. Each agency collects data for a different purpose and has embedded epistemological lenses and institutional traditions. Mapping projects of bear habitat in the province conducted by the provincial Ministry of the Environment (MoE) and by the Sierra Club of Canada might create different – though related – ontologies to describe forest cover, bear habitat, and interactions between the environment and wildlife. Imagine both projects are undertaken by professional researchers with extensive local and scientific knowledge as well as a commitment to accuracy. Nevertheless, it is inevitable that the two teams would develop different classification systems and represent forest interaction differently. The MoE must balance environmental concerns with economic productivity in a province that depends on logging for economic sustainability. Its classification and interpretation of forest cover, clear cut land and the effect on bear habitats is likely to vary considerably from a purely environmental perspective. In the case of the Sierra Club, we might find greater attention paid to the negative effect of clear cuts on bear habitat, migration paths, and the salmon streams that feed the bears. Not only is it expected that the MoE and the

Sierra Club will choose to define different data attributes, but their characterization of interaction between digitally defined spatial objects will also vary significantly. In each case, the two organizations would likely develop distinctive digital datasets with unique relationships between entities. Both organizations prioritize and legitimate 'evidence' differently in order to support their different agendas. In other words, they will create two different ontologies – based on their conflicting epistemologies.

Divergent ontologies always point to different epistemologies (or ways of studying and knowing the world) and result in different spatial entities and relationships between them. The development of multiple, deviating ontologies to describe environmental habitats is a reminder of the fact that all data are constructed with a specific social agenda. Each ontology is manufactured to represent a particular understanding of the world that emerges from a specific epistemology. So all data that populate a GIS ontology are socially produced. Metadata are a way of proving background and context to a particular ontology so that subsequent users have a clearer picture of the particular agenda and context that influenced the development of the dataset. Understanding that different ontologies – based on different ways of knowing the world – populate digital environments is essential to integrating metadata – and qualitative research principles – within GIS.

WHAT ARE METADATA AND HOW DO THEY DESCRIBE ONTOLOGIES?

Metadata are literally data about data. Datasets contain unique fields and each field is usually represented as a truncated data field (e.g. %Female or Crowncls). Seldom are these fields completely self-explanatory outside the institutional culture in which they are used. Metadata (data about the data) are a way of translating the field names into terms that can be understood by a wide range of data users – whilst retaining archival information about how the terms were developed and interpreted. Metadata answer questions like: when were the data collected; what time period do they cover; what scale are the data applicable to; what projection was used; how were the data collected; what quality measures were taken; how were the data classified; and what mapping units are associated with the data? Historically the metadata associated with GIS data focused on the spatial features (e.g. map projection) rather than the non-spatial attributes (e.g. forest type). Typically in GIS, a metadata form is provided by software programs as illustrated in Figure 3.1.

Metadata are a relatively recent addition to GIS datasets. In the first several decades of spatial data use, details of the lineage, scale, and quality of data were less important as data were not routinely shared (Schuurman, 2004). Data users implicitly understood how attributes were interpreted in their unique institutional culture. In addition, most data were collected for specific purposes – such as to make land use maps or designate zoning areas – by a government agency or business (Schuurman, 2005). The data were proprietary and seldom shared. It was only as organizations and individuals began to realize the value of sharing secondary data that data about the data became important as a means to assess the compatibility among different datasets.

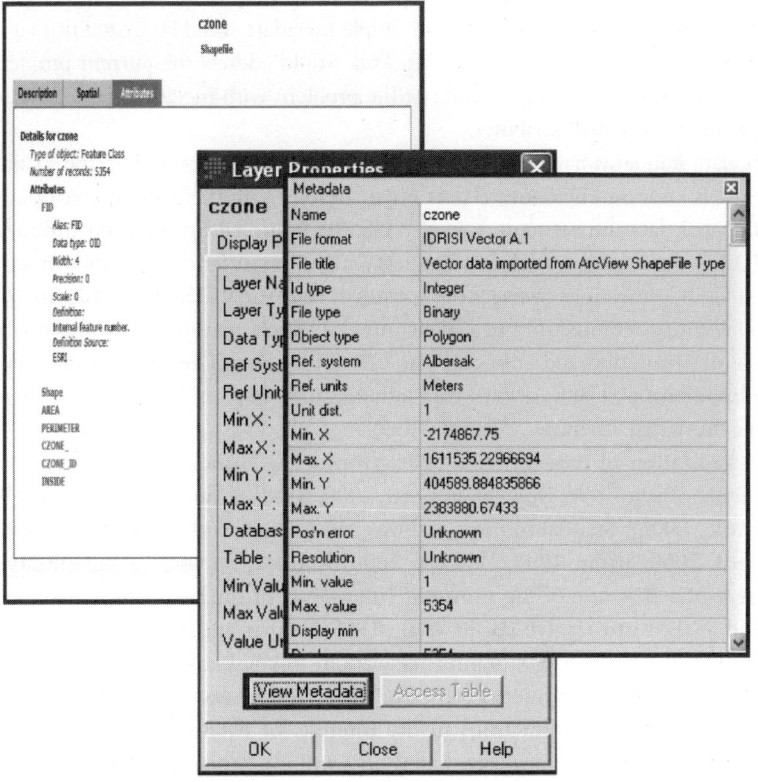

Figure 3.1 Two metadata forms. The left-hand side illustrates an ArcGIS metadata form, while the right-hand form is from IDRISI Andes. Note that metadata are collected exclusively for spatial attributes

Metadata usually describe *spatial* qualities of the dataset well. These attributes include projection, scale, and lineage. Data used in GIS must have a spatial component, so this is important. The vast majority of data fields, however, are non-spatial in nature – and lack detailed metadata to describe them. For instance, a perinatal data registry containing information before, during, and after births will contain the address of the mother and the location of the birth. These spatial data are necessary to enable mapping and analysis, but the vast majority of the fields are non-spatial and there are very few – if any – metadata to describe them.

Data sharing is essential as a means of broadening analyses. It allows users to map distributions across borders and between jurisdictions. Without metadata, it is very difficult to combine datasets that were collected by different organizations or individuals. Metadata are useful, for example, to create maps that compare the number of low-birthweight babies born in different countries or to compare pollutant volumes produced by coal burning in different jurisdictions. GIS software applications including Clark Labs'

IDRISI and ESRI ArcGIS have interfaces that support transfer and supplementation of metadata. Harlan Onsrud has gone a step further by proposing a 'geographic information commons' in which spatial data and simple metadata could be shared from a central web-based archive (Onsrud et al., 2004). This would address the current problem of a lack of availability for metadata – but not the problem with metadata sparseness or their singular focus on spatial attributes.

Metadata standards have been developed in many countries to describe spatial data. Their use is endorsed by national data organizations with the goal of developing consistent spatial data infrastructures. In the US, the National Spatial Data Infrastructure (NSDI) has developed spatial data formats (http://www.fgdc.gov/nsdi/nsdi.html); in Canada GeoConnections oversees the Canadian Geospatial Data Infrastructure (CGDI) (http://www.geoconnections.org/Welcome.do). Such national organizations play a key role in promoting and enforcing the use of metadata. Though metadata standards are an important part of a national data infrastructure, they are limited as their emphasis remains on spatial attributes (Masser, 1999).

Standardization of data terms is often proposed as a means of ensuring semantic equivalence (Bishr, 1998; Bowker and Star, 2000; Brodaric and Gahegan, 2001; Winter and Nittle, 2003). Standardization is, however, fraught with difficulties (Bibby and Shepherd, 2000; Bishr, 1998; Bowker, 2000). Indeed, semantic standardization has been recognized as one of the most difficult aspects of interoperability (or sharing of data and analysis) to resolve (Bishr et al., 1999; Bittner and Edwards, 2001; Brodeur et al., 2003; Kottman, 1999; Schuurman, 2005; Vckovski, 1999). In the absence of standardization, semantic integration of terms across databases is used to merge multiple databases across jurisdictions in the interests of data sharing and comparison. Integration aims to merge datasets based on common semantic terms such as 'boreal forest' or 'Canadian shield'.

Interpretations of data are, however, dynamic across time and space. Multiple databases associated with similar themes but from different institutions have unique ontologies and each database has ontological diversity that is extremely difficult – if not impossible – to unravel (Bowker, 2000). As a result it is very difficult to create semantic compatibility among databases from different contexts (Devogele et al., 1998; Laurini, 1998; Sheth, 1999). Obstacles include the difficulty of updating name changes and exporting meaning between databases – even when category names are standardized (Bowker and Star, 2000). The problem centers around the incontrovertible fact that attribute names are not equivalent to attribute meanings (Harvey, 1997; 1999a; 1999b; Harvey and Chrisman, 1998). Moreover, classification systems are unstable over time because the meanings of field names change implicitly (Bowker and Star, 2000; Harvey and Chrisman, 1998) as taxonomic regimes are updated to reflect new scientific understanding. The same classification system often contains terms with different meanings in multiple jurisdictions. Health data registries, for example, frequently use common terms such as 'blood antibodies' or 'diabetes' but these fields are defined differently in different jurisdictions (Schuurman and Leszczynski, 2006). Even when data dictionaries are used to create equivalencies, they remain unreliable because the name is not the same as the institutional meaning (Bowker and Star, 2000).

Semantic data integration is the process of linking individual fields between multiple datasets. It involves making assumptions about equivalence between two terms that might be alike but not necessarily equivalent. For instance, the word 'case' refers to the same general concept in the context of both police records of crime incidents and legal suits. The meanings are, however, sufficiently different that a naive integration of two such datasets could generate uncertainty in a subsequent analysis. This process of semantic data integration remains challenging because we have few well-established means of supplying information about non-spatial attributes between different organizations (Bishr, 1997).

Ironically, when interoperability between systems was first proposed and research initiated, scholars believed that semantic interoperability would be the simplest to solve as it involved only language (Bishr, 1998; Harvey, 1999a; Vckovski, 1999). The problem of language has, however, plagued researchers as words are intractable with hidden implicit meaning (Schuurman, 2002). Smith and Mark (2001) illustrate the difficulty of semantic data integration by pointing out that even with a set of criteria for what geographical objects fit the definition of 'mountain', many examples are lost. The trouble is that language is dynamic, contextual, and contingent on an existing framework. In other words, language is slippery; while computing, by contrast, requires and relies upon strict categories.

In GIScience, considerable progress has been made in developing technical solutions for sharing semantic data over the last decade (Bishr, 1997; Bishr et al., 1999; Harvey, 1999a; 2003; Kottman, 1999; Laurini, 1998; Vckovski, 1999). Research efforts include emphasis on context, semantic networks (Brodeur et al., 2003), data dictionaries (Agarwal, 2005), semantic proximity (Kavouras et al., 2005), automated mediators (Abel et al., 1998; Devogele et al., 1998) and semantic reference systems (Kuhn, 2003). These approaches are alike in that they frequently require that institutions restructure their data at considerable cost (e.g. from relational to object oriented databases). Most agencies and businesses are unlikely to undertake such reorganization with its attendant expense (Schuurman, 2002). What is needed instead is a means of sharing deep contextual information about non-spatial fields – for the purposes of sharing data and retaining archival information about semantic meaning – that can be piggybacked on to extant datasets. Contextual or ontological information about data is a means of ensuring that data are not misinterpreted in new contexts (Kashyap and Sheth, 1996).

ONTOLOGY-BASED METADATA: WHAT DO THEY LOOK LIKE?

Ontology-based metadata are a metadata solution that includes contextual information or deep meaning with the database terms (Schuurman and Leszczynski, 2006). This information must be embedded with the data so that it can be referred to for archival, comparison, and integration purposes. The advantage of ontology-based metadata is that they do not require novel computing or data structures and can be instituted after the data have been collected and are in use. In effect, ontology-based metadata extend the scope of existing spatial metadata to include the context of non-spatial attributes. They are a record of the epistemological legacy of the data. In other

Table 3.1 Eight fields used to capture ontology-based metadata. The fields were chosen to encapsulate key elements that describe instrumental, storage model, institutional culture, and policy context

Field	Description
Sampling methodologies	What sampling methods were used? Are the data a complete sample (e.g. class list) or a partial sample (e.g. Canada Census)?
Definition of variable terms	What is the standard field definition? Are naming conventions in place?
Measurement system	How were the data measured? Using what instrumentation (e.g. survey or GPS)?
Taxonomic system	What classification system was used?
Data model	What data model was used? Is there a history of using different data models?
Collection rationale	What was the original motivation for data collection (e.g. government statistics or business survey)? Why were the data collected?
Policy constraints	Were there legal constraints or influences on the data collection or classification?
Anecdotes	Are there relevant additional comments that could help users understand the data?

words, the metadata provide vital information about the rationale and agenda of the collectors that can be used by subsequent or parallel users to determine if the data are appropriate for their project.

The goal of ontology-based metadata is to include information about the sociological, political, and technical influences that bear on data. In order for ontology-based metadata to be adopted, however, relative simplicity of implementation must be assured. Many sophisticated systems for achieving more logical and manageable data structures have been abandoned because of the complexity of implementing them in businesses and government agencies (Schuurman, 2005). Moreover, it is compulsory that existing relational databases remain intact as experience has indicated the reluctance of institutions to abandon the massive infrastructure associated with relational databases (Schuurman, 2002). In the absence of usability and easy implementation, ontology-based metadata tables would be rejected as too cumbersome for already extended data stewards to incorporate into information management systems. So ontology-based metadata need to capture the elusive qualities of context – implicit and otherwise – as well as information about measurement techniques and data models in a manner that is palatable to data stewards. To that end, we developed eight sample fields that capture context while using a structure that is compatible with computing. Table 3.1 illustrates the eight fields developed to express non-spatial metadata.[1]

Each of the eight fields is a pointer to deep, embedded context associated with data that are not normally revealed to the user. The sampling methodology, for instance, is usually implicit. Even when using national census data, we are not aware of whether a complete or partial sample was used. In Canada, for example, only 20% of the population is asked to fill out the long survey for Canada Census. In the United States, efforts are made to do a complete census in which every individual is enumerated. Foreign users of Canada Census data may not necessarily understand that the data are based on an incomplete sample – or that its statistical reliability is (ironically) greater because of the sampling method. The second field is a standard variable definition. In most cases, this field is already available and linked to the dataset. The measurement system field is a critical means of assessing whether historical information is compatible with present-day data in terms of measurement accuracy. In the past, address data or general directions such as 'four miles east on Highway One' may have been used to locate spatial entities. By contrast, contemporary data may be collected using high-accuracy GPS. The two location fields may not be compatible in terms of resolution. The fourth field specifies the classification or taxonomic system used to categorize the data. This is critical information when multiple datasets use similar semantics that vary in meaning and schematics. For example, 'crown closure' may appear in two forest cover classification systems but have two different definitions. Likewise, the field name, crown closure, may be differently placed in an object hierarchy. Each of these distinctions can be made manifest if the taxonomic/classification system is transparent. The fifth element of ontology-based metadata is the data model. This concerns the organizational structure of the data – which may be relational or object oriented. The sixth component is collection rationale. Bowker (1996) points out that much about the hidden institutional values associated with data can be determined by understanding the collection rationale. For example, bear habitat data collected by the Ministry of the Environment will differ from those collected by the Sierra Club. For the former the data are collected to maximize available cutting blocks, while for the latter the data are intended to illustrate the need for forest preservation in the interests of bear survival. The collection rationale, in other words, addresses the issue of whose interest the data are collected in, with the recognition that each dataset is informed by a specific agenda.

These contextual metadata fields are not written in stone. They are subject to revision and customization depending on the perspective of the developer and possible future users. For instance, a feminist researcher working with breast cancer survivors and their histories might include other contextual information such as where the interviews were conducted, and who asked the questions. There are many ways of making the extended metadata more appropriate for a given study. A pertinent issue, however, is that these metadata are still collected as fields and they must travel with the data they describe. They must be somewhat succinct and portable. They will never qualify as a qualitative study. They exist as indicators for data users who might want to understand data context in the interests of integration or representation of multiple ontologies. Ontology-based metadata constitute a window into the social production of data and the underlying epistemologies of data providers.

COLLECTING ONTOLOGY-BASED METADATA

Creation of fields for ontology-based metadata is not enough. A mechanism is required to collect those data. The author has introduced a methodology to gather deep contextual information about non-spatial attributes. *Database ethnographies* are a means of collecting metadata about datasets from data stewards. They build upon a rich tradition of ethnography developed in the social sciences but extend and transform the methodology to a cyborg context.

Increasingly, data are the only description of the world that is available to us. As consumers, we are personally represented in databases as the sum of our consumer patterns, credit history, citizenship, date of birth, and any recorded deviant behavior. Most of us would agree that this is a skeletal representation of our lives and person, yet many decisions are made based on this framework. Likewise, datasets – whether they represent the environment, hospital data, or molecular composition – are all minimal representations of exceedingly complex phenomena. Data stewards are, however, one source of information about the nature of the data. Semi-structured interviews with data stewards and other key persons involved in developing the data can reveal the Kind of information illustrated in Table 3.2.

The metadata shown in Table 3.1 can be collected from in-house systems managers and information technology (IT) specialists and subsequently used to guide data integration and comparison. As web-based applications for sharing data increase, there is the potential for communities who have a shared interest in a particular dataset to share ontology-based metadata on the web in a wiki format. This would take advantage of the open and collaborative nature of the web. In such a scenario, the original data collectors would leave contextual information in a shared web repository. Later users could update the shared metadata and explain particular characteristics of the data from their perspective. This would greatly enhance the dimensionality of existing datasets. At present, however, a version of metadata sharing is occurring in some disciplines in a semi-formal fashion to allow the use of multiple near-but-not-equivalent datasets, as explained in the examples below.

CONTEXTUALIZING DATA: TWO EXAMPLES

Land cover data are collected through a wide range of techniques ranging from remote sensing to airborne photography to field surveillance. The last of these permits the accumulation of very high-quality data for a small area and is time intensive; by contrast, remote sensing is a means of collecting data for very large areas with a minimum of cost. In addition, remotely sensed data can be secured for a time series, allowing comparison of the same area chronologically. Remotely sensed data, unlike airborne data, are usually flown for a wide variety of applications and the data are seldom classified or categorized for precisely the purpose they may be put to. There are, for instance, advantages and limitations to a wide variety of remotely sensed land cover data for developing model carbon cycle models.

Forests have a carbon cycle in which all the carbon atoms in the atmosphere, plants, soils, and animals circulate. Plants absorb CO_2 from the atmosphere and discharge oxygen back into the atmosphere. Likewise, there is an exchange of CO_2 between the oceans and the atmosphere. The dissolved CO_2 in the oceans is used by marine biota in photosynthesis. Each of these CO_2 swaps influences climate and ultimately climate change. In order to model carbon cycles or other changes to forest cover on a global scale, it is necessary to have data for the entire world from the same approximate time periods.

Table 3.2 Ontology-based metadata for hypertension in pregnancy for the British Columbia Perinatal Database Registry (BCPDR)

Field	Details
Definition of variable terms	The British Columbia Reproductive Care Program (BCRCP) defines hypertension as a BP of at least 140/90 mmHg read on at least two instances during pregnancy prior to delivery. The scheme differentiates between *pregnancy-induced hypertension*, defined as a rise in systolic blood pressure of ≥ 30 mmHg and a rise in diastolic pressure of ≥ 15 mmHg when these changes occur on two occasions at least 6 hours apart; and *other cause of hypertension*, which excludes renal disease and is used to codify hypertension that arises during delivery but did not present earlier antepartum. The latter is also termed *transient hypertension of pregnancy*
Measurement specification	Measurement is based on blood pressure thresholds; these are provided under 'Definition of variable terms'
Classification system	As of 1 April 2004, consistent with ICD-10-CM. To analyze trends over time, an in-house conversion tool is used to move between statistics coded using ICD-9-CM and ICD-10-CM
Data model	MS Excel
Collection rationale	To isolate onset of disease by differentiating between hypertension complicating pregnancy and hypertension complicating delivery
Policy constraints	As outlined under 'Classification system', the BCPDR maintains data in both ICD-9-CM and ICD-10-CM. When the BCRCP migrated to ICD-10-CM, several codes did not convert properly, with hypertension being pre-eminent amongst these problematic fields. Although ICD-10 is much more specific than ICD-9, conversion resulted in a significant loss of data resolution or granularity as ICD-10 does not capture severity of disease equally. Women previously determined to be severely hypertensive according to ICD-9 were not so by way of the more recent ICD-10 codes
Anecdotes	While the database differentiates hypertension complicating pregnancy from hypertension complicating delivery alone, it does not explicitly separate chronic disease from that idiopathic to pregnancy, save for instances of renal disease. There are discrepancies in how physicians report hypertension, with many clinicians not detailing etiology of disease. In other words, it is often clear that the mother is hypertensive, but the cause or nature of disease remains unlisted

Currently there are three global land cover datasets widely available: NOAA-AVHRR; TERRA-MODIS; and SPOT-VEGETATION. The chief differences between these datasets revolve around spectral properties (reflective characteristics) and resolution, image processing algorithms, temporal period, classification (legend), and validation procedures (Jung et al., 2006). Each of these characteristics has an impact on the usefulness of the final product in a given research or development context. The user frequently remains unaware of the complex influences of the spectral properties of the sensor, its resolution, or the algorithms used to process the data. Instead, they use only the classification system on which to base decisions about overlay and comparison of the different land cover datasets.

In order to compare different periods and geographical areas using combinations of the three main land cover datasets, the user requires a means of mapping similar classes between the different legends (e.g. ontologies). Jung et al. developed an innovative classification of the three land cover datasets by sorting each category from the three individual classification systems based on *functional* land cover. This new classification system included categories such as urban, trees, grasses, wetlands, and snow. This represented a great simplification of the original datasets but one that could pragmatically be used to model carbon cycles around the globe using a combination of the three chief land cover datasets. They were able to achieve this, however, by understanding in great detail the breakdown of categories and the various components such as resolution, processing algorithms, and validations methods that were used to determine the categories. In effect, they were using *implicit* ontology-based metadata. In other words, the authors used the information they had on the scientific epistemology and assumptions of the different data providers to develop a fuller and better contextualized picture of what each data field really meant. They then used this context to create a framework for integrating data from the three datasets. Implicit in their reconstruction and data integration was the assumption that each dataset was the product of a different knowledge system but that the ontologies referred back to the same spatial objects on the earth's surface.

The metadata that they gathered for this project were the key to semantic integration of three disparate land cover datasets. Those metadata are, however, not a fundamental part of these three datasets and had to be explicitly gathered for this project. The advantage of widespread ontology-based metadata lies in their potential to facilitate data integration and comparison across near-but-not-equivalent datasets from diverse sources for a range of users.

In the next example, ontology-based metadata were developed as part of a pilot project to compare two different perinatal datasets. In this case, *explicit* contextual metadata were developed for each dataset and are bundled with the data. Including explicit ontology-based metadata provides a means for current and future researchers to assess the applicability of the data as well as to use the metadata for semantic integration.

Pregnancy-induced hypertension is a complex term as it encompasses two distinct diseases. The first instance occurs when *chronic* or *pre-existing hypertension* exists (often undetected) prior to pregnancy, and the second occurs when the disease is idiopathic to increased heart stress associated with pregnancy. In the latter case, pregnancy-induced hypertension resolves spontaneously with the removal of the placenta at birth. In the case of pre-existing disease, hypertension continues after the birth of the child. Definitional issues are rife as the initial symptoms of chronic or pre-existing hypertension often coincide with the pregnancy, making a specific and precise diagnosis difficult. A number of hypertensive disorders associated with pregnancy have been identified in medical literature (Cnossen et al., 2006; Duda, 1996; Ihle et al., 1987; Roberts, 1981; Xiong et al., 1999; Zhou et al., 1997). However, medical convention maintains that a blood pressure (BP) reading of >140/90 mmHg at < 20 weeks gestation is often a marker for an underlying hypertensive disorder. Excluding pre-existing disease, two cohorts of women experiencing hypertension in pregnancy can further be differentiated: women whose hypertension presents in late pregnancy and remains static towards term with no adverse outcomes in mother or fetus, and those women where hypertension develops at some point in pregnancy and progresses aggressively, often presenting adverse renal and placental effects (Symonds, 1980). Distinguishing between these two hypertensive-in-pregnancy populations presents yet another definitional problem as this differentiation can only be determined post-partum when symptoms can be observed to abate or persist (Symonds, 1980). Both pathologies can be referred to as gestational hypertension or pregnancy-induced hypertension in different medical data registries.

Explicit ontology-based metadata were gathered through interviews with data stewards at the British Columbia Perinatal Database Registry (PDR). The simplicity of the ontology metadata fields enables context to be gathered without arduous investigation while providing pertinent archival and integration/comparison information. Table 3.2 contains explicit ontology-based metadata for the British Columbia PDR.

When semantic integration is conducted for the purposes of national or international comparisons of rates, it is assumed that 'equivalent' terms have the same meaning. These terms are, however, used and interpreted divergently in different medical settings. Our research revealed that terms like 'stillbirth' and 'fetal death', and 'gestational hypertension' and 'pregnancy-induced hypertension', are used differently across institutional settings – and in database hierarchies. Ontology-based metadata are a means of recording those differences. In the process of enhancing the meaning of non-spatial attribute data, metadata offer the promise of increasing the reliability of results generated by spatial analysis. The GIS analyst, in this instance, could map clusters of each type of hypertension with more accuracy based on understanding how the two databases correspond.

CONCLUSION: CONTEXT MATTERS

Data are not in themselves authoritative or even true. They are the product of context and are always socially produced. Their collection for a certain purpose at a particular time in a specific place may be implicit but it is not transparent to subsequent users. There is no exact or reliable method for extracting that context or encoding it – but efforts to understand context (ontologies) and provide a reliable, transportable record of it constitute important first steps toward acknowledging that all data and all maps are perspectival views of the world with their own value system and agenda.

Ontology-based metadata constitute a means of capturing an albeit limited perspective on the context that surrounds the data used for spatial analysis in GIS. The fields that populate ontology-based metadata can be designed by the user community or they can use those suggested here. One could argue that metadata do not fix problems; rather they generate new ones in that the metadata themselves now require context. The response to this lies in the fact that data are always imperfect but there remains the opportunity to optimize them. The ultimate goal of ontology-based metadata is to ensure that data are used not as the 'truth' about the world, but rather as information that is contextual and contingent. By ensuring that future and parallel users understand the strengths and limitations of the data, we can achieve more reliable analyses and better cartographic communication.

In this chapter, I have explained divergent views of ontology. Philosophers and information scientists differ in their understanding of ontology: philosophers interpret ontology as a singular reality, while computer scientists acknowledge multiple ontologies that must each be represented in computational terms. Metadata are a means of encapsulating descriptive information about ontologies that are represented digitally. Interviews with data stewards, users, and collectors are a means of gathering ontology-based metadata that can be stored in simple relational database tables linked to datasets. Ontology-based metadata are a powerful means of enhancing spatial data so that they carry with them a record of their history and their implicit meaning. This information is a way of making GIS a tool for producing multiple rather than singular narratives.

ACKNOWLEDGEMENTS

Support for this research was provided by the Canadian Institutes of Health Research (CIHR) grant no. 116338. I would also like to thank Agnieszka Leszczynski for her invaluable help with creating the figure and the tables.

NOTE

1 An earlier version of Table 3.1 was published in Schuurman and Leszczynski (2006).

REFERENCES

Abel, D.J., Ooi, B.C., Tan, Kian-Lee and Tan, S.H. (1998) 'Towards integrated geographical information processing', *International Journal of Geographical Information Science,* 12 (4): 353–71.

Agarwal, Pragya (2005) 'Ontological considerations in GIScience', *International Journal of Geographical Information Science,* 19 (5): 501–36.

Bibby, Peter and Shepherd, John (2000) 'GIS, land use, and representation', *Environment and Planning B: Planning and Design,* 27: 583–98.

Bishr, Y.A. (1997) 'Semantic aspects of interoperable GIS'. PhD dissertation, Spatial Information, International Institute for Aerospace Survey and Earth Sciences (ITC), Enschede, The Netherlands.

Bishr, Y.A. (1998) 'Overcoming the semantic and other barriers to GIS interoperability', *International Journal of Geographical Information Science,* 12 (4): 299–314.

Bishr, Y.A., Pundt, H., Kuhn, W. and Radwan, M. (1999) 'Probing the concept of information communities – a first step toward semantic interoperability', in M.F. Goodchild, M. Egenhofer, R. Fegeas and C. Kottman (eds), *Interoperating Geographic Information Systems*. Boston: Kluwer. pp. 55–69.

Bittner, Thomas and Edwards, Geoffrey (2001) 'Towards an ontology for geomatics', *Geomatica*, 55 (4): 475–90.

Bowker, G (1996) 'The history of information infrastructures: the case of the International Classification of Diseases', *Information Processing and Management*, 32 (1): 49–61.

Bowker, G. (2000) 'Mapping biodiversity', *International Journal of Geographical Information Science*, 14 (8): 739–54.

Bowker, G. and Star, S.L. (2000) *Sorting Things Out: Classification and Its Consequences*. Cambridge, MA: MIT Press.

Brodaric, Boyan and Gahegan, Mark (2001) 'Learning geoscience categories *in situ:* implications for geographic knowledge representation', paper presented at the Ninth ACM International Symposium on Advances in Geographic Information Systems, Atlanta, GA.

Brodeur, Jean, Bedard, Yvan, Edwards, Geoffrey and Moulin, Bernard (2003) 'Revisiting the concept of geospatial data interoperability within the scope of the human communication processes', *Transactions in GIS*, 7 (2): 243–65.

Cnossen, Jeltsje S., van der Post, Joris A.M., Mol, Ben W. J., Khan, Khalid S., Meads, Catherine A. and ter Riet, Gerben (2006) 'Prediction of pre-eclampsia: a protocol for systematic reviews of test accuracy', *BMC Pregnancy and Childbirth*, 6 (29).

Devogele, T., Parent, C. and Spaccapietra, S. (1998) 'On spatial database integration', *International Journal of Geographical Information Science*, 12 (4): 335–52.

Duda, Jennifer (1996) 'Preeclampsia, still an enigma', *The Western Journal of Medicine*, 164 (4): 315–20.

Fonseca, Frederico T., Egenhofer, Max J., Agouris, Peggy and Camara, Gilberto (2002) 'Using ontologies for integrated information systems', *Transactions in GIS*, 6 (3): 231–57.

Frank, Andrew U. (2001) 'Tiers of ontology and consistency constraints in geographical information systems', *International Journal of Geographical Information Science*, 15 (7): 667–78.

Gregory, D. (1994) 'Ontology', in R.J. Johnston, D. Gregory and D.M. Smith (eds), *The Dictionary of Human Geography*. Oxford: Blackwell.

Gregory, D. (2000a) 'Ontology', in R.J. Johnston, D. Gregory, G. Pratt and M. Watts (eds), *The Dictionary of Human Geography*. Oxford: Blackwell.

Gregory, D. (2000b) 'Realism', in R.J. Johnston, D. Gregory, G. Pratt and M. Watts (eds), *The Dictionary of Human Geography*. Oxford: Blackwell.

Gruber, T. (1995) 'Toward principles for the design of ontologies used for knowledge sharing', *International Journal of Human–Computer Studies*, 43: 907–28.

Harvey, F. (1997) 'From geographic holism to geographic information systems', *The Professional Geographer*, 49: 77–85.

Harvey, F. (1999a) 'Designing for interoperability: overcoming semantic differences', in M.F. Goodchild, M. Egenhofer, R. Fegeas and C. Kottman (eds), *Interoperating Geographic Information Systems*. Boston: Kluwer.

Harvey, F. (1999b) 'What does interoperability mean?', paper presented at The Annual Meeting of the Association of American Geographers, Honolulu, HI.

Harvey, F. and Chrisman, N.R. (1998) 'Boundary objects and the social construction of GIS technology', *Environment and Planning A*, 30: 1683–94.

Harvey, F. (2003) 'Developing geographic information infrastructures for local government: the role of Trust', *The Canadian Geographer*, 47 (1): 28–36.

Ihle, B.U., Long, P. and Oats, J. (1987) 'Early onset pre-eclampsia: recognition of underlying renal disease', *British Medical Journal*, 294 (6564): 79–81.

Jung, M., Henkel, K., Herold, M. and Churkina, G. (2006) 'Exploiting synergies of global land cover products for carbon cycle modeling', *Remote Sensing of Environment*, 101: 544–53.

Kashyap, V. and Sheth, A. (1996) 'Semantic and schematic similarities between database objects: a context-based approach', *The VLDB Journal*, 5: 276–304.

Kavouras, M., Kokla, M. and Tomai, E. (2005) 'Comparing categories among geographic ontologies', *Computers & Geosciences*, 31 (2): 145–54.

Kottman, C. (1999) 'The Open GIS Consortium and progress toward interoperability in GIS', in M.F. Goodchild, M. Egenhofer, R. Fegeas and C. Kottman (eds), *Interoperating Geographic Information Systems*. Boston: Kluwer. pp. 39–54.

Kuhn, Werner (2003) 'Semantic reference systems', *International Journal of Geographical Information Science*, 17 (5): 405–09.

Laurini, R. (1998) 'Spatial multi-database topological continuity and indexing: a step towards seamless GIS data interoperability', *Geographical Information Science*, 12 (4): 373–402.

Masser, I. (1999) 'All shapes and sizes: the first generation of national spatial data infrastructures', *International Journal of Geographical Information Science*, 13 (1): 67–84.

Onsrud, H., Camara, G., Campbell, J., Chakravarthy, N.S. (2004) 'Public Commons of geographic data: research and development challenges', in M.J. Egenhofer, C. Freska and H.J. Miller (eds), *Geographic Information Science*. Berlin! Springer-Verlag. Lecture notes in computer science #3234.

Roberts, J.M. (1981) 'Preeclampsia and eclampsia', *The Western Journal of Medicine*, 135 (1): 34–43.

Schuurman, N. (2002) 'Flexible standardization: making interoperability accessible to agencies with limited resources', *Cartography and Geographic Information Science*, 29 (4): 343–53.

Schuurman, N. (2004) *GIS: A Short Introduction*. Oxford: Blackwell.

Schuurman, N. (2005) 'Social perspectives on semantic interoperability: constraints to geographical knowledge from a database perspective', *Cartographica*, 40 (4): 47–61.

Schuurman, N. and Leszczynski, A. (2006) 'Ontology-based metadata', *Transactions in Geographic Information Science*, 10 (5): 709–26.

Sheth, A.P. (1999) 'Changing focus on interoperability in information systems: from system, syntax, structure to semantics', in M.F. Goodchild, M. Egenhofer, R. Fegeas and C. Kottman (eds), *Interoperating Geographic Information Systems*. Boston: Kluwer.

Smith, Barry and Mark, David M. (2001) 'Geographical categories: an ontological investigation', *International Journal of Geographical Information Science*, 15 (7): 591–612.

Symonds, E.M. (1980) 'Aetiology of pre-eclampsia: a review', *Journal of the Royal Society of Medicine*, 73 (12): 871–5.

Vckovski, Andrej (1999) 'Interoperability and spatial information theory', in M.F. Goodchild, M. Egenhofer, R. Fegeas and C. Kottman (eds), *Interoperating Geographic Information Systems*. Boston: Kluwer.

Winter, Stephan (2001) 'Ontology: buzzword or paradigm shift in GI science?', *International Journal of Geographical Information Science*, 15 (7): 587–90.

Winter, Stephan and Nittle, Silvia (2003) 'Formal information modelling for standardisation in the spatial domain', *International Journal of Geographical Information Science*, 17 (8): 721–41.

Xiong, Xu, Hayes, Damon, Demianczuk, Nestor, Olson, David, Davidge, Sandra, Newburn-Cook, Christine and Saunders, L. Duncan (1999) 'Impact of pregnancy-induced hypertension on fetal growth', *American Journal of Obstetrics and Gynecology*, 180 (1, pt 1): 207–13.

Zhou, Yan, Damsky, Caroline H. and Fisher, Susan J. (1997) 'Preeclampsia is associated with failure of human cytotrophoblasts to mimic a vascular adhesion phenotype: one cause of defective endovascular invasion in this syndrome?', *Journal of Clinical Investigation*, 99 (9): 2152–64.

4

MULTIPLE REPRESENTATIONS, SIGNIFICATIONS AND EPISTEMOLOGIES IN COMMUNITY-BASED GIS

Sarah Elwood

ABSTRACT

This chapter shows how the everyday GIS practices of grassroots groups can contribute to a qualitative GIS repertoire. Such qualitative GIS practices need not involve new software innovations, but rather may emerge through mixed methods multi-epistemology engagements with GIS that rely creatively upon its visualization capabilities and its capacity to incorporate multiple forms of information. Drawing examples from participatory research conducted with two Chicago, Illinois community-based organizations, I illustrate how their GIS use integrates qualitative and quantitative forms of information, represents their neighborhood as social space and material space, and engages GIS as both representative and constitutive of identities and meanings that are bound to particular spaces. These examples illustrate an approach to qualitative GIS characterized by epistemological and methodological flexibility and multiplicity, and most significantly, one that may be practiced by novice GIS users.

INTRODUCTION

> People talk about, 'Division Street, that's a horrible place.' But then other people talk about, 'Paseo Boricua is such a wonderful place.' They're the same physical space, but psychologically, they're two different environments, two different realities. And the problem, is, how do you show [both]. [Rey, 2005][1]

In this statement, a community organizer describes some of the representational and communicative challenges of his organization's use of GIS in neighborhood planning, problem solving and redevelopment activities. He refers to multiple meanings that may be associated with a single place, and the way that invoking place names activates these meanings. Division Street is a major commercial corridor on Chicago's northwest side, a portion of which crosses through the Humboldt Park area and has been portrayed in the media as a dangerous, declining place. A coalition of non-profit organizations called the Humboldt Park Empowerment Partnership (HPEP) has fostered an enclave of

Puerto Rican businesses and community agencies along a portion of the street. They call this section Paseo Boricua[2] and seek to rework not just material conditions but meanings associated with this place. As some community development organizations in the Humboldt Park area have begun to use GIS, they have developed strategies for illustrating these multiple meanings, in spite of the challenges alluded to by the organizer quoted above. Their GIS practices integrate qualitative and quantitative forms of information to associate a range of meanings with neighborhood spaces, in ways that are strategically shifted to support different priorities or goals.

These engagements with GIS contribute to a growing range of mixed methods or qualitative approaches to GIS. As other chapters in this collection illustrate, qualitative GIS encompasses a wide range of interventions with 'GIS as we know it' (Sheppard, 2001: 546), from reworking representational or analytical capabilities of software, to using GIS for inductive interpretive approaches to knowledge production, to weaving GIS into a mixed methods research design (Dennis, 2006; Knigge and Cope, 2006; Kwan, 2002a; 2007; Pain et al., 2006; Pavlovskaya, 2006). Shared commitments in qualitative GIS include integrating both qualitative and quantitative forms of reasoning or ways of representing spatial conditions and relationships; incorporating multiple ways of knowing; and using analysis techniques from both quantitative and qualitative traditions. Drawing evidence from extended participatory research with two community organizations, I will show that some grassroots groups make significant contributions to this qualitative GIS repertoire through their everyday practices with GIS. Their efforts illustrate that GIS is far more open for this modification than has been recognized. Even novice GIS users with limited resources can and do remake GIS in ways that challenge the epistemological and representational limits that have been assumed in much of the academic literature.

REWORKING GIS: PGIS, GIS/2 AND OTHER RECONSTRUCTIONS

Critical perspectives on GIS in the mid 1990s raised a number of concerns about GIS, including the extent to which it could incorporate diverse forms of spatial knowledge and promote multiple epistemologies (Pickles, 1995; Sheppard, 1995). In the years following, researchers and practitioners working in participatory or public participation GIS (PPGIS), feminist GIS, qualitative GIS, and critical GIS have sought to challenge some of these characterizations. Among other things, feminist and qualitative GIS have sought to disengage some of the linkages that continue to inform some geographers' assumptions about the technology: GIS as a quantitative method, GIS as rooted in a positivist epistemology, or GIS as necessarily supporting rationalist knowledge claims or technological determinism. Schuurman (2002a; 2002b) and Kwan (2002a), for instance, have illustrated how GIS is used within realist or pragmatist epistemologies, arguing against claims of positivism. Pavlovskaya (2006; and this volume) argues that GIS applications are rarely distinguishable as either quantitative or qualitative and usually engage elements of both. She argues further that visualization, often a critical mechanism in realizing the social and political power of GIS, may engage in the same intermingling of quantitative and qualitative elements.

After unharnessing GIS from a limited set of research designs, epistemologies or analysis techniques, approaches framed as qualitative GIS have sought to reinvent these linkages in several ways. Some researchers have developed GIS-based methods that integrate qualitative and quantitative approaches to analysis and theorization, such as Pain et al.'s (2006) 'qualified GIS' or Knigge and Cope's (2006) 'grounded visualization'. Others have developed techniques for incorporating non-cartographic information into GIS databases or for including multiple or non-quantitative attributes of a space or spatial object. Examples include Kwan (2002b) and Kwan and Lee's (2004) use of three-dimensional, animated, and artistic techniques to represent emotional geographies; Sieber's (2004) use of markup languages and tags to encode alternative attributes; and the multimedia GIS approaches that enable linking oral narratives, photographs, and other forms of knowledge to a GIS (Al-Kodmany, 2000; Shiffer, 1998; Weiner and Harris, 2003).

These contributions have greatly expanded accounts of the forms of knowledge, representation, and analysis that are possible with GIS. But most such efforts to support qualitative or mixed methods approaches have emerged from academic researchers' efforts to reconfigure GIS or recast its intellectual and analytic role within a research project. If, as Schuurman (2002a) and Kwan (2002a) have argued, GIS is open for rewriting through the critical agency of GIS users, it is imperative to also understand how other kinds of users are remaking GIS through their practices. Participatory GIS (PGIS) initiatives provide an important window on creative reworking of GIS. PGIS refers to the development of geospatial technologies that is initiated and directed by participants, often in participatory development processes in the global South (Rambaldi et al., 2004). It emphasizes collaborative processes for developing spatial data, and inclusion of multiple realities, diverse perceptions of space and place. The PGIS literature is replete with examples of how GIS might be used with non-cartographic or qualitative ways of representing and communicating spatial knowledge, including oral narrative, sketch mapping, performance, or collaborative collection of 'field data' (Kyem, 2004; Rambaldi, 2005; Rambaldi et al., 2006; Tripathi and Bhattarya, 2004; Weiner and Harris, 2003; Williams and Dunn, 2003).[3]

The predominant focus in the published literature on PGIS has been upon its social and political processes, illustrating how GIS, spatial data, and maps produce and negotiate politics and power relations, or how they can be used to foster participatory decision making processes. Far less attention has been given to the way that grassroots groups, community organizations, activists, and ordinary citizens working in participatory contexts use GIS in ways that reconfigure conventional understandings of the kinds of knowledge that GIS may be used to produce and share.[4] There are relatively few accounts of how the sorts of practices promoted and studied in PGIS may produce GIS in new ways. I take up this approach here, examining how the GIS practices of grassroots users contribute to a reworking of GIS toward a qualitative or mixed methods GIS. 'Grassroots GIS users' refers to the kinds of individuals and organizations that tend to be involved in PGIS initiatives: smaller NGOs, activist groups, community organizations. Characterizing these groups as grassroots GIS users is admittedly vague, but given the large number that have adopted GIS, they can no longer be considered new or non-traditional GIS users. The range of activities in which they are involved varies such that merely characterizing them as local or community based is also problematic.

Examining the impacts of grassroots GIS practices upon GIS, rather than primarily upon social and political relationships, reveals some of the same methodological and epistemological blurring and boundary crossing that are part of academic researchers' engagements with qualitative GIS. A mixed methods, multi-epistemology GIS practice is forged in the efforts of some grassroots GIS users, in this case, local community development organizations. Drawing from ongoing participatory research with two community development organizations in Chicago, in the following sections I will illustrate how their GIS practices advance multiple delineations of neighborhood boundaries; represent the neighborhood as both a material and a social space; rely on both qualitative and quantitative approaches for binding meanings and identities to these spaces; and engage GIS as simultaneously representative and constitutive of identities, knowledge, and places and their characteristics. I use this case to illustrate how grassroots groups create a mixed methods, multi-epistemology GIS practice, not to suggest their efforts to do so are unique from those of other grassroots groups. This case illustrates how PGIS practice can contribute to a remaking of GIS by bridging many of the separations or oppositions that have been assumed in other characterizations of GIS. This remaking is not wrought by altering software or hardware, or through embedding multimedia forms of spatial knowledge. Rather it relies on the representational flexibility inherent in existing forms of the technology, creatively mixing and shifting representations, epistemologies, and signification strategies.

THE HUMBOLDT PARK COMMUNITY GIS PROJECT

The Near Northwest Neighborhood Network (NNNN) and the West Humboldt Park Family and Community Development Council are non-profit community development and organizing groups that work in an area of Chicago's northwest side. Both groups address a range of challenges, including loss of affordable housing, poor housing conditions, higher rates of crime and unemployment, and inadequate health and public education resources. Their activities include strategic planning; providing information and advocacy for housing and small business development; sponsorship of affordable housing projects; and community organizing activities such as block clubs, safety programs, or youth and family support services.

While NNNN is assumed to work in Humboldt Park, and the Development Council in West Humboldt Park, these places are difficult to define precisely. The City of Chicago designates boundaries of Humboldt Park as one of its 77 community areas, an official administrative designation roughly analogous to a neighborhood. Community areas do not correspond to electoral wards or to community organizations' and residents' understanding of neighborhood boundaries. In practice, residents, community organizations, and even local governmental officials themselves delineate the boundaries of Humboldt Park in a variety of ways, many of which are different than the boundaries of the Humboldt Park Community Area.[5] West Humboldt Park has no such official designation as a community area. As a place name, West Humboldt Park emerged as a signifier of demographic change in the western portion of Humboldt

Park as African Americans moved to the area, alongside the primarily Latino residents of Humboldt Park. Humboldt Park also functions as a racial and ethnic signifier, referencing this place as Latino, but more specifically referencing its identity as the historic center of Chicago's Puerto Rican community.

NNNN and the Development Council began using GIS in 2003 as part of a collaborative research project involving me, staff, residents, university-based research assistants, and students in my undergraduate and graduate GIS courses. The goals of this project include examining how community organizations use GIS-based spatial data and maps to try to influence planning, problem solving, and service delivery, and to understand the social and political implications of these GIS practices. Staff members at both organizations have learned to use GIS software, and in collaboration with residents and the university students, created a spatial data library that supports their GIS activities. The data library includes information gathered from a variety of government sources, such as the US Census or the City of Chicago's GIS Division, as well as information collected in the Humboldt Park area by staff, students, and residents. The groups use the GIS and data library to create maps illustrating neighborhood needs and conditions, their own activities and accomplishments in the area, and future plans for neighborhood redevelopment efforts. These maps are used in funding proposals; their own strategic planning, community organizing, and coordination with other community groups; and efforts to advocate for or against particular decisions with policy makers.

As I have discussed elsewhere (Elwood, 2006a; 2006b), the community organizations do not treat these outputs from their GIS use simply as fixed thematic maps. Rather, they employ them as flexible cartographic texts that may be used to narrate shifting characterizations of a place, in support of many different agendas or projects. This strategy involves interpretive framing of a map, accomplished through titles or other text on a map, oral discussion of it in a presentation, or written text in an accompanying document, such as a grant proposal. These frames may be shifted to reinterpret a single map to advance several different narratives, in one instance framing a map as evidence of unequal distribution of local government resources, while in another framing it as evidence of organization accomplishments in bringing needed services to the area. These flexible spatial narratives are an example of what Del Casino and Hanna (2005) articulate as representational practice. They argue that maps are productive not just in the representative moment of their creation, or in the discourses advanced through the visible text that appears on a map, but in the spaces and meanings that are produced when maps are reinterpreted and reframed for specific agendas. Interrogating maps not just as representations but as representational practices, they argue, disrupts many of the binaries through which we think of maps and map making, such as map user versus map maker.

NNNN and the Development Council's creation and ongoing reworking of GIS-based spatial narratives can be read as representational practice in much the same way. The GIS practices of these two organizations disrupt several of the binary assumptions about GIS with respect to representation, signification, and epistemology. Their GIS practices integrate multiple representations of space and place, adopt multiple approaches to signifying identities and meanings, and engage GIS from multiple epistemological positions with respect to the truth claims it can advance. I develop these arguments

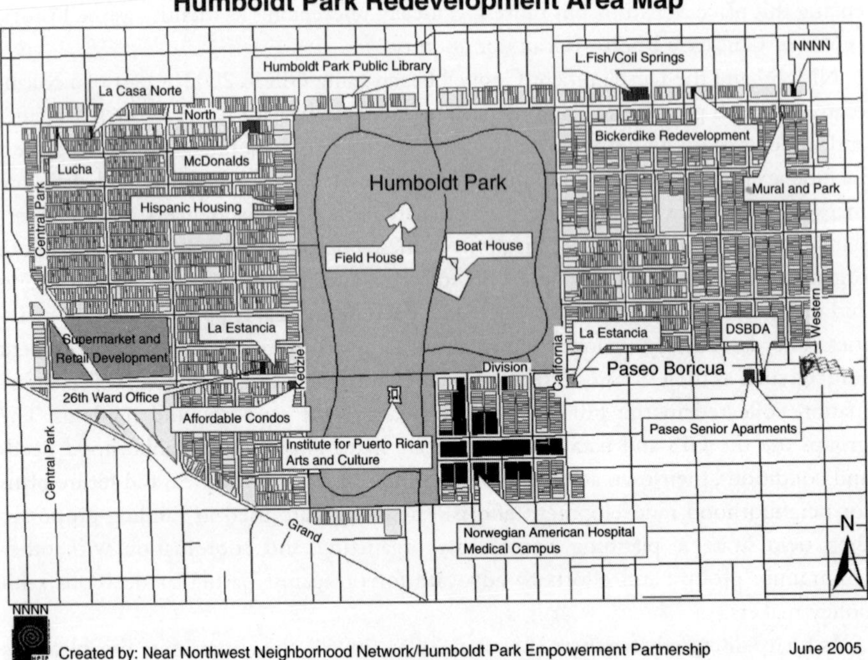

Figure 4.1 NNNN's redevelopment area map

from participant observation of the groups' GIS activities and development of spatial data; analysis of community newspapers, presentations, and other archival materials in which they include their GIS-based data or maps; and semi-structured interviews with staff and volunteers. Through interpretive analysis of this evidence, I explore the discursive and symbolic composition of these maps, the reasons why the map makers create particular compositions, and how they engage these cartographic texts to support specific agendas or priorities.

WHO/WHERE/WHAT IS (GREATER) (WEST) HUMBOLDT PARK? GIS PRACTICES AT THE GRASSROOTS

In this section, I examine NNNN and the Development Council's GIS practices with reference to three maps that they have used widely and shared with other individuals and organizations.[6] Figure 4.1, the redevelopment map, created by NNNN, shows affordable housing resources, services for residents, and community cultural resources that HPEP organizations helped to develop. It is widely used in NNNN's organizing efforts with residents, in grant proposals, and in walking tours or strategic planning meetings with policy makers. Figure 4.2, the census map, was also created by NNNN,

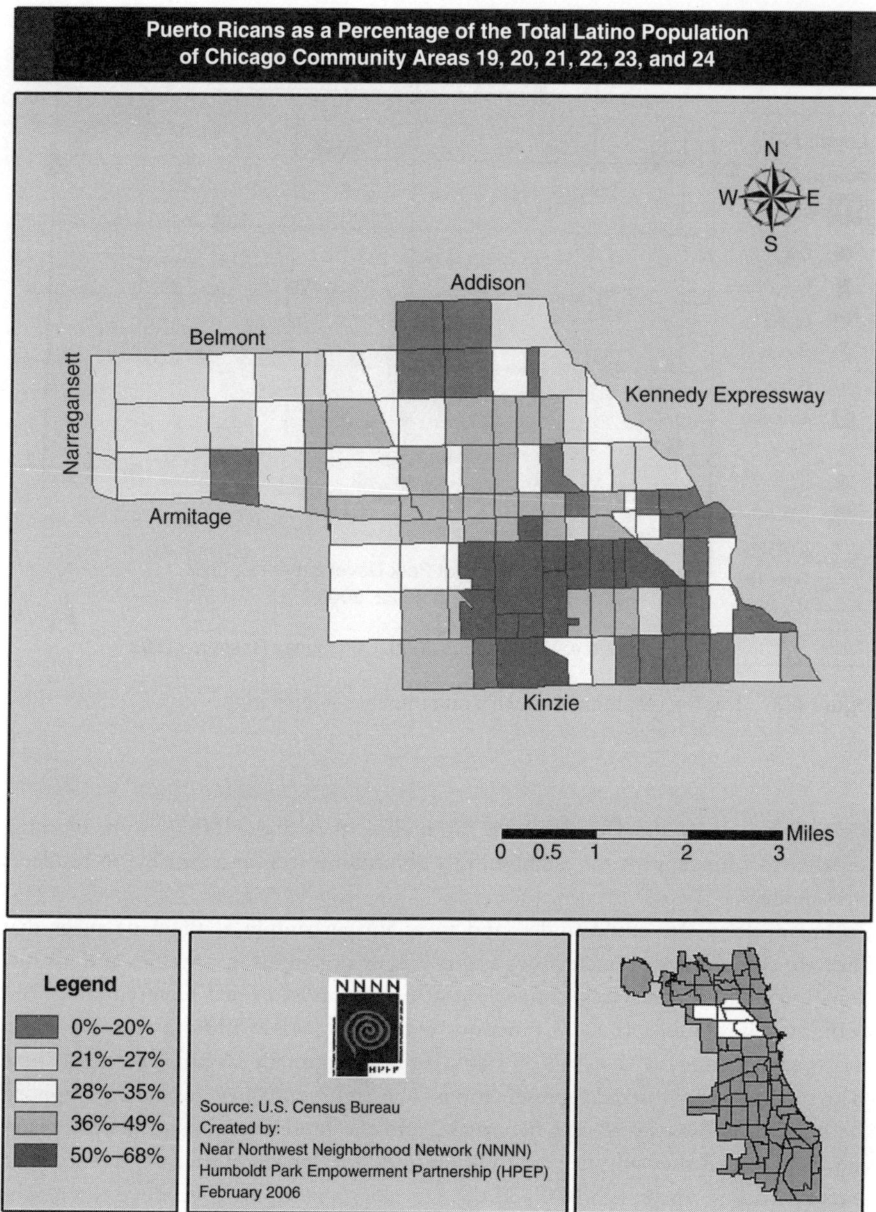

Figure 4.2 NNNN's census map

using census data to show Puerto Ricans as a percentage of the total Latino population in the Humboldt Park Community Area and several nearby community areas. Figure 4.3, the community services map, was created by the Development Council and shows public services in the area of a prospective community center. This map was one of

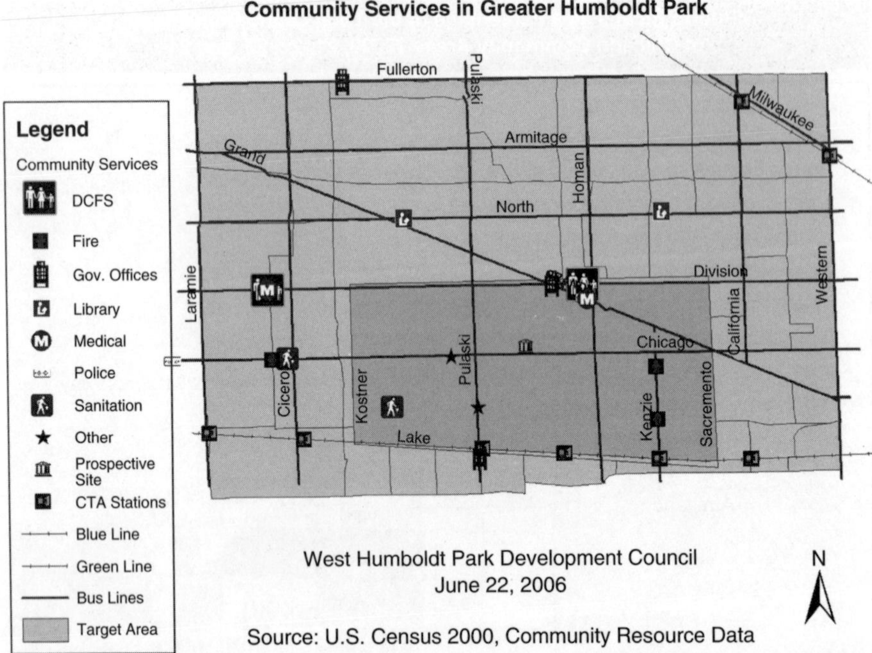

Figure 4.3 The Development Council's community services map

a series created by the Development Council as they and NNNN were working together to gain support for establishing a community center accessible to residents of both areas.

On one level, we could understand these images simply as thematic maps that illustrate community characteristics and needs, or organization activities and institutional presence and services, conceiving of each map as a fixed representation. But examining how multiple maps function together as part of shifting representations and practices suggests that NNNN and the Development Council perform these maps as anything but fixed representations. The arguments they are used to support, the types of knowledge claims they make, and the representations of people, place, and identity that they advance are multiple and fluid. Through their representational practice of these maps, NNNN and the Development Council produce the neighborhood as material *and* as imagined, bind neighborhood identities to these spaces through quantitative *and* qualitative approaches, and engage GIS as both representative *and* constitutive of community knowledge and identities. In so doing, they blur boundaries between methods, representational practices, and epistemologies that have typically been understood as separate, producing GIS as far more flexible and open than has often been recognized.

Neighborhood as fixed and flexible, material and imagined

Since 2003, NNNN and the Development Council have produced several hundred maps, which assign many different boundaries and names to the area in which they operate. The area that their staff members refer to as 'our community' or 'our neighborhood' may be delineated with different boundaries that cover a wide array of locations on the northwest side of Chicago. The demarcated space may be identified as Humboldt Park, East Humboldt Park, West Humboldt Park, or Greater Humboldt Park. The community services map (Figure 4.3), for instance, identifies a territory termed Greater Humboldt Park that covers nearly the entire northwest and central west portions of the city. In contrast, the redevelopment area map (Figure 4.1) identifies as Humboldt Park a very small area north and east of what is identified as Greater Humboldt Park in the community services map. In these shifting delineations, the term 'Humboldt Park' rarely refers to the officially designated boundaries of the Humboldt Park Community Area. Hence, it is not this 'official' boundary that is being changed. Rather, a variety of other boundaries are drawn on the map to represent the Humboldt Park community in ways that support particular projects or agendas. As noted earlier, even local government officials use the identifier 'Humboldt Park' to identify spaces that do not correspond to the officially recognized community area.

The shifting boundaries and place names in these two maps and others created by NNNN and the Development Council function to produce a neighborhood in two very different ways. First, the bounded territories included in these maps establish a neighborhood as a fixed object, so that this object can be connected with different meanings, activities, or agendas.[7] Representing a neighborhood as a fixed territory enables linking this object to the mission and legitimacy of the two organizations, in ways that are critical to their ability to seek resources, policy change, or the support of residents for their work. Simply put, a community development organization must have a place on whose behalf it operates. If there is to be a West Humboldt Park Family and Community Development Council, there must be a West Humboldt Park. Or, as one of the NNNN organizers explains:

> We're a Housing Resource Center through the Chicago Department of Housing, so we have a contract through them to provide technical assistance to residents here in our service area. [Alonso, 2004]

Queried further, he explains that the service area is Humboldt Park, the area on whose behalf NNNN works. This explanation illustrates how, in the contemporary politics of urban governance and community development, defining a territory is essential in legitimizing an organization's prerogative to advocate for policies or decisions on behalf of the people who live there.

However, at the same time, the supposedly fixed objects of Humboldt Park, West Humboldt Park, and other places are constantly in flux as both groups use their GIS software to revise boundaries and names. These rebounding and renaming efforts enable the organizations to connect with a wider range of meanings, projects, and potential allies. For instance, in the community services map (Figure 4.3), the use of Greater Humboldt Park marks this map as the product of a collaborative effort by both NNNN and the Development Council to bring a community center to the area. With Humboldt Park and West Humboldt Park so strongly delineated as the service areas or operating territories for the two organizations, neither place name is sufficient to delineate the space of an initiative sponsored by both groups. Thus, shifting boundaries and place names creates places open for new kinds of engagements or collaborations.

This fluidity of boundaries and place names across different maps illustrates that any presentation of Humboldt Park, West Humboldt Park, or Greater Humboldt Park as a clearly delineated fixed spatial object is a strategic performance toward some end. In so far as it is a defined territory that enables an organization like NNNN to legitimize or support its work in that area, Humboldt Park as a neighborhood has some element of materiality. But simultaneously, it exists as a collectively imagined space. As one of the organizers from NNNN explains:

> [Humboldt Park] really exists more within people's minds. If you look at Humboldt Park, the community area defined by the city of Chicago, that's for the most part what we call West Humboldt Park. But a lot of people who live in what technically would be West Town consider themselves to live in Humboldt Park, or they might distinguish it as East Humboldt Park. [Andrea, 2006]

Another NNNN staff member, referring to how the place names Humboldt Park and West Humboldt Park operate as racial signifiers, similarly illustrates these places as collectively imagined:

> We have all these [residents] who want to stay in the Greater Humboldt Park area. But some people say, 'Well, West Humboldt Park is not Humboldt Park', and that's because of seeing these places as Latino and Black. [Hector, 2006]

Here, the organizer describes a socially constructed distinction between the two places, based upon residents' assumptions about the people who live in those places.

In these examples we see NNNN and the Development Council using GIS to activate two different conceptualizations of neighborhood simultaneously – a neighborhood as a bounded material object, but also as an imagined space constructed through organizational agendas, residents' sense of place, and other social and political processes. In a single map delineating Humboldt Park, the neighborhood is produced as a fixed entity. This appearance of fixity can be politically advantageous for the community organizations, when they want their representation of a place or its characteristics to be understood by map readers as incontrovertible, thus legitimizing the organizations' claims. One staff member explained this approach, recounting a colleague's defense when he was questioned about a map presented at a meeting with local government officials:

> They're like, 'Are you sure this is it?' And he's like, 'Yes, it's there. It's on the map.' [Teresa, 2004]

This strategic deployment of 'fixed' places stands in contrast to their persistent redrawing of boundaries, renaming of territories, and efforts to shift the meanings associated with these places. In these practices, a neighborhood is flexible and open for reworking to support many different initiatives, priorities, or meanings. This too can be advantageous for the community organizations, since it opens the door to strategic deployment of carefully considered 'truths' to audiences with different priorities, agendas, or experiences.

GIS is central to advancing both productions – the cartographic representation of a supposedly fixed territory and its continued digital and social reinvention. Activists and community organizations tend to have deep expertise in producing and negotiating meaning in ways that advance their causes, but adopting GIS does more than simply provide them a new vehicle for existing practices. Rather, because of the unique potential of GIS to support this strategic engagement with flexible and fixed representations, GIS becomes an inextricable part of new practices of meaning making.

Binding identity to place through quantitative and qualitative approaches

Another way in which NNNN and the Development Council's integrated GIS practice bridges often-separated methodologies or modes of representation is through their use of quantitative and qualitative approaches to associating identities or meanings with places. For instance, as previously discussed, Humboldt Park is understood across the city as the historic center of Chicago's Puerto Rican community. For NNNN and other organizations in HPEP, continuing to assert Humboldt Park as Puerto Rican is important for reasons that include cultural pride, anti-gentrification strategies, historical preservation, or even sustaining community organizations. Many of the maps made at NNNN over the past four years work to reinforce this identification. These maps rely upon multiple forms of representation, evidence, or measurement in their effort to associate a Puerto Rican identity with the spaces of Humboldt Park.

More generally, most of the maps created by NNNN and the Development Council are produced by compiling text, symbols, and one or more data layers from the community organizations' spatial data library. They tend not to use traditional GIS techniques for spatial analysis such as buffering or spatial overlay. Rather, they communicate a particular message or bind meanings to space through cartographic representation of different forms of data. Some of these data, such as measurements of housing or population characteristics from the US Census, are clearly quantitative. Some data, such as photographs of buildings along the major business corridors in the neighborhood, are more qualitative representations of local conditions, illustrating these conditions through a photographic image rather than, say, symbolization of the percentage of boarded storefronts they have observed along this corridor in their field data collection. Other data are difficult to definitively characterize as either quantitative or qualitative, such as information compiled on the location, names, sponsoring agencies, and costs of housing and economic development projects in their service areas.

The representation of these quantitative or qualitative forms of data on their maps nearly always relies upon classic cartographic symbolization, using visual variables such as color, pattern, or shape. But different forms of information – quantitative or qualitative – may be used to develop similar claims. That is, the same characterization of a neighborhood space may be sought using representations of both quantitative and qualitative information. The census map (Figure 4.2) and the redevelopment area map (Figure 4.1), for instance, rely upon qualitative and quantitative representations of Humboldt Park as Puerto Rican. The census map makes this association by illustrating that the majority of the Latino population in the census tracts comprising the eastern portion of the Humboldt Park Community Area is Puerto Rican. In this approach, the linking of 'Humboldt Park' and 'Puerto Rican' occurs through the representation of a quantitative measurement of the demographic characteristics of the population. The redevelopment area map makes this association on very different terms, based upon the presence of Puerto Rican cultural sites in the area, or the presence of projects and institutions whose names reference linkages to Puerto Rico. Among other significations, this map includes and labels Paseo Boricua (the enclave of Puerto Rican institutions described in the opening paragraph of this chapter), the Paseo Senior Apartments, the Institute for Puerto Rican Arts and Culture, and La Estancia. La Estancia is a mixed-use affordable housing project whose name translates as 'The Stand', a direct reference to the efforts of NNNN and other organizations to combat the displacement of lower-income Latino residents by gentrification. Other visual strategies in the map also stake this space as Puerto Rican, such as the icon of the Puerto Rican flag that appears on the eastern edge. The approaches adopted here for claiming Humboldt Park as Puerto Rican rely upon representing the presence of Puerto Rican institutions and the historical and cultural significance of the area for Puerto Ricans in Chicago, rather than upon measuring the demographic identification of residents.

Neither approach can be framed as wholly quantitative or qualitative. However, the census map does rely upon a quantitative measurement in its portrayal of Humboldt Park as Puerto Rican. In contrast, the redevelopment area map relies upon a distinctly non-quantitative approach for signaling Humboldt Park as Puerto Rican. It uses text, icon, and other visual symbols that stand in for different components on which this claim is based: cultural enclave spaces, institutions, and place names that reference Puerto Rico, and accomplishments by organizations understood as primarily Puerto Rican. Some quantitative measurement might have been used, such as the percentage of businesses along Paseo Boricua. Rather, they use the nominal presence of certain businesses, housing, or organizations to make the case for the area as Puerto Rican.

Whether these claims are advanced through qualitative or quantitative approaches, these maps are part of a politics that engages two very different epistemological assumptions simultaneously. In their work on behalf of the neighborhood, NNNN staff and residents clearly treat Humboldt Park or its identification as Puerto Rican as social and political constructs, created to support various claims or agendas, *and* simultaneously as though they have some objective or demonstrable materiality. For instance, speaking about why the census map and the redevelopment map make the case for Humboldt Park as Puerto Rican, one of the community organizers says:

> If you look at Humboldt Park, it was majority Puerto Rican and [now it is] Mexican … There is a deliberate attempt to create and sustain a [Puerto Rican] cultural enclave, because of the threat of gentrification. Through time, it is inclusive of other people … because we're mostly living side-by-side, and we have similar struggles. [Hector, 2006]

In this description, Hector treats Humboldt Park and its portrayal as Puerto Rican as a constructed performance produced for a specific project, the organization's anti-gentrification efforts.

In other instances, staff and residents frame the place and its characteristics from a very different set of initial assumptions. At a community meeting about affordable housing, a resident who has worked for NNNN and other community organizations objected to the seeming invention of a new neighborhood through the formation of a group calling itself the West Bucktown Neighborhood Association. Bucktown is the name usually given to a rapidly gentrifying area of the near northwest side. To applause from the staff and residents assembled around the table, she said:

> Where is West Bucktown, anyway? There is no such thing as West Bucktown. We all know that is Humboldt Park! [Liliana, 2004]

In this statement, especially juxtaposed against the angry suggestion that West Bucktown does not exist, there is a clear assumption that Humboldt Park *does* exist somehow, in a way that is somehow *real*, in contrast to the construct of West Bucktown. Both of these examples center upon NNNN's efforts to combat gentrification in the eastern portion of its service area. But the first operates from constructivist assumptions about places and their characteristics, while the second operates on the basis of realist assumptions. More importantly for the arguments taken up in this chapter, the organization's integrated GIS practices support both approaches.

GIS as constitutive and representative: engaging local epistemologies

The practices described above illustrate that these community organizations approach GIS as representative and as constitutive of community knowledge and identities. On one hand, the maps they create tend to reflect predominantly held notions about a place and its characteristics, representing Humboldt Park as Puerto Rican, for instance, or West Humboldt Park as African American. Simultaneously, these maps are also constitutive. As they are used by residents at community meetings, shared with local government officials, or discussed by staff at other non-profits, the maps reinforce residents' own ideas about their neighborhoods, as well as outsiders' understandings of these places. This constitutive effect is clearly evident in the continued insistence of many residents that Humboldt Park *is* Puerto Rican, even when maps such as the census map (Figure 4.2) suggest that in 2000 only a small portion of the neighborhood had more than 50% of the population that identified as Puerto Rican. Of course, the response of residents to a census questionnaire is not the only way in which Humboldt Park could *be* Puerto Rican, as evidenced by the effort in the redevelopment map (Figure 4.1) to make this case based upon the presence of Puerto Rican institutions in the neighborhood. But the point is that these characterizations do not only represent identifications of place, they produce and reinforce them.

As before, these representative and constitutive engagements with GIS involve a tension between fixed and flexible productions of places and their characteristics. So, for instance, each individual map of Humboldt Park, West Humboldt Park, or Greater

Humboldt Park frames a supposedly stable entity with clearly defined boundaries. But using the GIS to produce these places iteratively and dynamically with widely ranging areas, boundaries, and names intentionally constitutes and reconstitutes these places in flexible ways for specific agendas and to support different characterizations or claims about them. That is, in their production and use of these maps, people dynamically engage with Humboldt Park as some kind of extant spatial reality *and* as collectively imagined and socially produced. Claims about Humboldt Park as Puerto Rican advanced in the census map and the redevelopment area map can operate similarly, depending on how a map is framed or interpreted for a given audience or presentation. A staff member can present either of these maps as representing some tangibly evident Puerto Rican presence in the Humboldt Park area; or, as in Hector's reflection (quoted above), the maps may be a self-aware production of situated places and meaning, as in efforts to produce Humboldt Park as Puerto Rican as part of an anti-gentrification agenda.

These constitutive/representative engagements with GIS are linked to the complicated politics these organizations are creating, framed in the previous section as a politics rooted in approaching place and its characteristics from both constructivist and realist perspectives. An additional piece of this politics is their intentional reliance upon the discursive 'expert power' of GIS and the assumptions that other individuals and institutions may make about reality when it is represented in GIS-based data and maps. Both organizations actively engage these assumptions to try to add greater weight to the claims they advance through their GIS-based data and maps. Staff members have observed that many of the powerful actors and institutions whom they seek to influence treat GIS-based data and maps as illustrations of what is real or true about a place, and as evidence of an expert (and therefore, legitimate) portrayal of that place. The community organizer's statement quoted earlier in this chapter ('Yes, it's there. It's on the map.') illustrates intentional deployment of this expert power when the organizer invokes the map as incontrovertible evidence of the correctness of his portrayal of land use along a major industrial corridor in the Humboldt Park area.

I have found further evidence supporting the community organizations' claim that other actors are likely to treat GIS-based data and maps as accurate representations of local conditions. One of the elected officials in the Humboldt Park area, for instance, when asked about NNNN and the Development Council's use of GIS, said:

> If you present that kind of information to people, they're more likely to believe what it is that you're talking about. [Juan, 2005]

Because GIS-based data and maps carry this discursive power as expert and accurate, presenting them can facilitate the efforts of these organizations to insert their perspectives and priorities as authoritative representations of an actually existing reality. But as these groups operate in a political context characterized by ever-changing allies, resources, and accepted urban revitalization strategies, they also find it beneficial to engage with GIS from a flexible constructivist perspective.

These strategies are not without their limitations and potential problems. Most significantly, the success of efforts to shore up the claims made in a map, either through invoking the expert power of GIS or through careful strategic scripting of data and symbolization, rests strongly upon the interpretation of that map by readers. While the community map makers certainly see their efforts to bind meaning to place in their maps as influential, they cannot fully control readers' interpretations. Beyond the composition and symbolization of a map, the map makers may also try to influence interpretation through their presentation, in oral discussion of the map, or in surrounding text in a grant proposal, an organization newsletter, or a community newspaper article. These additional authorings have their own limitations. An oral presentation exists primarily in the moment, and the tangible artifact of the map might be separated from the surrounding text and later understood differently by a reader. Or the fixity of the material artifact of the map may perpetuate a characterization of the community that changes over time, or it may obscure conflicting views or alternative perspectives. Further, my account of the political significance and impacts of these strategies for the community organization is based on the staff members' and residents' representation of these strategies to me, whether explained in interviews or demonstrated in situations where I was acting as a participant observer. I have closely observed how local government officials have responded to maps presented by the two organizations, but have been unable to directly question these actors about their own understandings of the community maps or their response to the flexible/fluid spatial meanings advanced in them.

In sum, the integrated practices these groups have developed suggest that GIS may be engaged from multiple epistemological positions. Several scholars have already illustrated that GIS lends itself to more than one epistemology (Kwan, 2002a; Pavlovskaya, 2006; Schuurman, 2002a), but this case shows GIS being engaged in multiple epistemologies simultaneously. These different engagements with GIS function as what Longino (2002) terms a local epistemology.[8] Local epistemologies are validated based upon their appropriateness for understanding a given situation or for fostering change under certain conditions, rather than upon some external validation criteria. In this case, we see how GIS is open for reworking to support different local epistemologies, and the adeptness of some novice GIS users in performing this reworking.

This case also shows that these flexible epistemologies may be created without necessarily altering GIS software or using non-Cartesian forms of spatial knowledge. There are of course strong examples of how software-level changes or use of non-Cartesian spatial knowledge can expand the epistemological diversity of GIS (Chmara-Huff, 2006; Sieber, 2004; see also Jung, this volume). But this case shows that GIS is open for flexible epistemological reworking without these interventions – an observation that is critical to appreciating the significance of GIS for grassroots groups. They are more likely to be novice users without the resources to reconfigure hardware, software, or databases. These novice GIS users do have a wealth of experience crafting complex representations of people and place, and working these representations to their own advantage amidst shifting political opportunity structures, and this expertise can be used to craft GIS practices that are multiple, flexible, and influential.

CONCLUSION

The GIS practices of these grassroots GIS users constitute one kind of qualitative GIS, one that is characterized by ontological, epistemological, and methodological flexibility and multiplicity. The practices described here engage only some of many possible quantitative–qualitative intersections that could comprise qualitative GIS, as seen in the range of approaches demonstrated by the other chapters in this collection. But they share in common with other qualitative GIS approaches a practice of blurring or complicating some of the binary distinctions that have been assumed to characterize GIS practice: fixed versus flexible definitions of spatial objects and their attributes; qualitative versus quantitative approaches for signifying spatial characteristics or meanings; and realist versus constructivist epistemologies. What these examples of grassroots GIS practice illustrate is that these productive representational, methodological, and epistemological integrations need not necessarily alter conventional GIS software or data structures. The uses of GIS described here are quite conventional and quite simple, if we focus solely upon the operations performed with the GIS software or analyze each individual visual output created. However, examining the creation and use of these GIS-based spatial data and maps, particularly the diversity of knowledge claims they are used to advance, reveals a more complex range of representations, epistemologies, and ontologies being leveraged through these GIS practices.

This case also speaks to the close intertwining of socio-political and technological processes that is inherent in GIS development and use. Even as NNNN and the Development Council are reworking GIS through their strategic engagement with fixed and flexible representations, or their iterative reconstruction of spatial meanings to speak to multiple audiences or priorities, so too are these engagements reworking them. That is, this cultivation of fluid construction of meaning as a political practice cannot be separated from the inherent representational flexibility of GIS as a technology for working with spatial knowledge. They are co-productive of another, and it is these co-productive socio-political and technological processes that qualitative GIS attempts to engage.

NOTES

1 All quotations are drawn from interviews and participant observation with research participants. In keeping with our confidentiality agreements, they are identified only by a pseudonym.

2 'Boricua' is used by some Puerto Ricans to refer to themselves ('Yo soy Boricua'or 'I am Puerto Rican') or to refer to the island itself. Both uses draw on the indigenous Taíno name for the island prior to Spanish colonization.

3 I have included only a few representative citations here. For a far more extensive bibliography of PGIS references, see the Integrated Approaches for Participatory Avenues in Development website (http://www.iapad.org).

4 Notable exceptions include Sieber (2000) and Weiner and Harris (2003).

5 For instance, an HPEP volunteer offered an anecdote about the Chicago Housing Authority defining Humboldt Park to include nearly all of the northwest side, incorporating large portions of four adjacent community areas: Austin, West Garfield Park, East Garfield Park, and North Lawndale. This redrawing of Humboldt Park, the volunteer argued, was intended to support the CHA's efforts to rehouse displaced public housing residents from central west side neighborhoods, while still ensuring that this relocation met their own policy requirement that residents be resettled in their original community of residence.

6 The maps were originally produced in color, but must be reproduced in black and white here. Where shading differences are difficult to discern in the reproduced images, I provide narrative description of the original image.

7 Dennis (2006) provides an excellent discussion of how visual representation of objects may then be 'bound' to discourses.

8 See Leitner and Sheppard (2003) for a useful discussion of Longino's concept of a local epistemology and its potential significance in critical urban geography research.

REFERENCES

Al-Kodmany, K. (2000) 'Extending geographic information systems (GIS) to meet neighborhood planning needs: recent developments in the work of the University of Illinois at Chicago', *The URISA Journal*, 12 (3): 19–37.

Chmara-Huff, F. (2006) 'A critical cultural landscape of the Pahrump Band of Southern Paiute'. Unpublished masters thesis, University of Arizona, Tucson.

Del Casino, V. and Hanna, S. (2005) 'Beyond the "binaries": a methodological intervention for interrogating maps as representational practices', *ACME*, 4 (1): 34–56.

Dennis, S. (2006) 'Prospects for qualitative GIS at the intersection of youth development and participatory urban planning', *Environment and Planning A*, 38 (11): 2039–54.

Elwood, S. (2006a) 'Beyond cooptation or resistance: urban spatial politics, community organizations, and GIS-based spatial narratives', *Annals of the Association of American Geographers*, 96 (2): 323–41.

Elwood, S. (2006b) 'Negotiating knowledge production: the everyday inclusions, exclusions, and contradictions of participatory GIS research', *The Professional Geographer*, 58 (2): 197–208.

Knigge, L. and Cope, M. (2006) 'Grounded visualization: integrating the analysis of qualitative and quantitative data through grounded theory and visualization', *Environment and Planning A*, 38 (11): 2021–37.

Kwan, M. (2002a) 'Is GIS for women? Reflections on the critical discourse in the 1990s', *Gender, Place and Culture*, 9 (3): 271–9.

Kwan, M. (2002b) 'Feminist visualization: re-envisioning GIS as a method in feminist geography research', *Annals of the Association of American Geographers*, 92 (4): 645–61.

Kwan, M. (2007) 'Affecting geospatial technologies: toward a feminist politics of emotion', *The Professional Geographer*, 59 (1): 22–34.

Kwan, M. and Lee, J. (2004) 'Geovisualization of human activity patterns using 3D GIS: a time-geographic approach', in M. Goodchild and D. Janelle (eds), *Spatially Integrated Social Science*. New York: Oxford University Press. pp. 48–66.

Kyem, P. (2004) 'Power, participation, and inflexible institutions: an examination of the challenges to community empowerment in participatory GIS applications', *Cartographica*, 38: 5–18.

Leitner, H. and Sheppard, E. (2003) 'Unbounding critical geographic research on cities: the 1990s and beyond', *Urban Geography*, 24 (6): 510–28.

Longino, H. (2002) *The Fate of Knowledge*. Princeton: Princeton University Press.

Miller, C. (2006) 'A beast in the field: the Google Maps mashup as GIS', *Cartographica*, 41 (3): 187–99.

Pain, R., MacFarlane, R., Turner, K. and Gill, S. (2006) '"When, where, if, and but": qualifying GIS and the effect of streetlighting on crime and fear', *Environment and Planning A*, 38 (11): 2055–74.

Pavlovskaya, M. (2006) 'Theorizing with GIS: a tool for critical geographies?', *Environment and Planning A*, 38 (11): 2003–20.

Pickles, J. (1995) 'Representations in an electronic age: geography, GIS, and democracy', in J. Pickles (ed.), *Ground Truth: The Social Implications of Geographic Information Systems*. New York: Guilford. pp. 1–30.

Rambaldi, G. (2005) 'Who owns the map legend?', *The URISA Journal*, 17 (1): 5–13.

Rambaldi, G., McCall, M., Weiner, D., Mbile, P. and Kyem, P. (2004) 'Participatory GIS', http://www.iapad.org/participatory_gis.htm.

Rambaldi, G., Kyem, K., McCall, M. and Weiner, D. (2006) 'Participatory spatial information management and communication in developing countries', *The Electronic Journal of Information Systems in Developing Countries*, 25 (1): 1–9.

Schuurman, N. (2002a) 'Reconciling realism and social constructivism in GIS', *ACME*, 1 (1): 73–90.

Schuurman, N. (2002b) 'Women and technology in geography: a cyborg manifesto', *The Canadian Geographer*, 46 (3): 258–65.

Sheppard, E. (1995) 'GIS and society: towards a research agenda', *Cartography and Geographic Information Systems*, 22: 5–16.

Sheppard, E. (2001) 'Quantitative geography: representations, practices, and possibilities', *Environment and Planning D: Society and Space*, 19 (2): 535–54.

Shiffer, M. (1998) 'Multimedia GIS for planning support and public discourse', *Cartography and Geographic Information Systems*, 25 (2): 89–94.

Sieber, R. (2000) 'Confronting the opposition: the social construction of geographical information systems in social movements', *International Journal of Geographic Information Systems*, 14 (8): 775–93.

Sieber, R. (2004) 'Rewiring for a GIS/2', *Cartographica*, 39 (1): 25–40.

Tripathi, N. and Bhattarya, S. (2004) 'Integrating indigenous knowledge and GIS for participatory natural resource management: state-of-the-practice', *The Electronic Journal of Information Systems in Developing Countries*, 17 (3): 1–13.

Weiner, D. and Harris, T. (2003) 'Community-integrated GIS for land reform in South Africa', *The URISA Journal (APAII)*, 15: 61–73.

Williams, C. and Dunn, C. (2003) 'GIS in participatory research: assessing the impacts of landmines on communities in northwest Cambodia', *Transactions in GIS*, 7 (3): 393–410.

5

GEOGRAPHIC INFORMATION TECHNOLOGIES, LOCAL KNOWLEDGE, AND CHANGE

Jon Corbett and Giacomo Rambaldi

ABSTRACT

Increasingly, local communities throughout the world are using a range of cartographic tools and technologies to depict the complex set of relationships between themselves and their territories. These range from community produced paper maps and the building of physical participatory three-dimensional models, through to the deployment of sophisticated geographic information systems (GIS). Collectively they are referred to as geographic information technologies (GIT). The thematic commonalities shared between these different tools and technologies are manifested through the application of processes which are typified by being initiated, guided, and realized at the community level, as well as a desire to communicate information about traditional landscapes and seascapes to decision makers as well as future generations. The *raison d'être* behind creating these products often rests on the assumption that the cartographic medium is a commonly understood and recognized visual language that is both effective and powerful in communicating this community–space relationship. GIT have the potential to be a medium that allows local communities to represent themselves spatially and thereby contribute to gaining recognition and inclusion in land and sea as well as natural resource claims, planning, and management. The production and use of community GIT can often have profound and at times unforeseen ramifications. From the process of creation through to their application and use, they have the capacity to impact social institutions within the community as well as wider relationships on social, cultural, and political levels. These ramifications and impacts to a large degree are determined by a series of enabling and disabling local, regional, and national environments. Drawing on case studies, research, and experiences from the Fiji Islands, Indonesia, Kenya, and the Philippines, this chapter will explore some of the social and political issues related to the creation and use of GIT in gathering and representing local knowledge in the struggle for some communities to gain local autonomy over traditional lands and development processes and safeguard their cultural heritage.

Figure 5.1 Ogiek elders are video-recorded while sharing their memories related to locations visualized on the three-dimensional map of their ancestral lands, Nessuit, Kenya, 2006

> We have learnt things about our land that we had forgotten.
>
> I felt happy to see the way our land has been represented, this will help our children to know their territory.
>
> Our rights that were hidden have now come to the light.

These short statements were written on cards and stuck on large sheets of kraft paper hanging on the wall of the village where elders had gathered to map the boundaries and resources of their ancestral territories. The mapping exercise represented a milestone in a long-running process initiated by the Ogiek indigenous peoples living in the Mau forest complex in Kenya to regain their lost lands. Assisted by intermediaries, the Ogiek organized and presented their spatial memories through the manufacturing of a 1: 10,000 scale, georeferenced, three-dimensional model (see Figure 5.1). The process of manufacturing the model, developing a mutually agreeable legend, and superimposing mental maps on the blank model lasted 11 days and involved 85 Ogiek elders and 37 youths representing 21 clans. Deep reflection and intense negotiation among members of different clans accompanied the process. Assisted by the model, elders would locate and articulate their spatial knowledge, often with great excitement. Each feature placed on the model, whether a pin, a string, or a smear of paint, captured memories that were supported by handwritten notes and audiovisual media These multiple sources and mixed media representing the tangible and the intangible heritage of the Ogiek people were later transposed into a GIS with the consent of the knowledge holders.[1]

This modeling process stimulated community cohesion, helped reclaim lost memories about the traditional ways of living as hunter-gatherers, facilitated intergenerational knowledge exchange, and raised awareness across generations and participating stakeholders about the critical status of the environment in terms of depleted forest cover and affected watershed functions. Community members concluded that they had a more holistic understanding of their social, cultural, and biophysical environments and that they realized the importance of working together towards a common goal. They further stated that they became aware of the value and potential authority of their spatial knowledge once it was collated, georeferenced, documented, and visualized. This example illustrates how participatory community mapping may be a kind of qualitative GIS by way of its richly interactive and reflective processes of negotiating and representing knowledge through diverse media, experiences, and ways of knowing. In this chapter we develop these linkages in more detail, characterizing processes and politics of participatory community mapping in a growing number of struggles for self-determination in the global South.

COMMUNITY MAPPING IN THE GLOBAL SOUTH

The view that 'maps … convey a sense of authority' (Alcorn, 2000: 1) has contributed to the 'premise that mappers engage in an unquestionably "scientific" or "objective" form of knowledge creation' (Harley, 1989) that ultimately represents the 'truth' and shapes the way that we understand the spatial world around us. However, this misconception has increasingly been questioned by academic discourse seeking to reveal the subjective and manipulative nature of geographic information and cartographic communication and questioning the objective and apolitical claims of the scientific model (Belyea, 1992; Crampton, 1995; Dahl, 1992; Harley, 1988; 1989; 1990; Harvey, 1990; Monmonier, 1991; Wood, 1992; Wright, 1942).

Representation of geographic information through the science of cartography is not neutral and is in no way separate from the broader power relations present in society (Livingstone, 1992). Since the inception of Cartesian map making, colonial and ruling powers have used maps as a tool to exert their claims over land (Edney, 1993; Wood, 1992). These claims have often been made to the detriment of societies already living on the land (for examples see Brody, 1981; Crawhall, 2001; Harris and Weiner, 1998; Peluso, 1995; Poole, 1995).

Despite forces that have served to exclude non-experts from map making, a growing number of local communities[2] and organizations associated and working together with communities (including development facilitators and technology intermediaries from non-governmental organizations, community-based organizations, universities, and development agencies) have begun to harness the potential power associated with maps for their own gain. These initiatives are commonly referred to as 'community mapping'. This is a map-making process that attempts to make the association between

Figure 5.2 Ogiek peoples using aerial images to locate their traditional lands, Nessuit, Kenya, 2005

land and local communities visible to outsiders by using the commonly understood and recognized language of cartography.

Community mapping projects have sprung up throughout the world – from Southeast Asia, through Central Asia, Africa, Europe, and North, South, and Central America, to the Pacific and Australasia (Chapin et al., 2005; Crawhall, 2001; Poole, 1995; Stan and Amiel, in press).[3] Many different types of communities have undertaken mapping projects, ranging from relatively prosperous groups to local communities and indigenous groups in the tropics.

Community maps often differ considerably from more mainstream maps in content, appearance, and methodology. They represent a socially or culturally distinct understanding of landscape and include information that is excluded from conventional maps, which usually represent the views of the dominant sectors of society. Community maps can pose alternatives to the languages and images of the existing power structures and become a medium of empowerment (Peluso, 1995). They have the potential to enable local communities to spatially represent themselves and their relationship to their local physical, socio-cultural, and biological environments (Figure 5.2). Yet concomitantly community maps have the potential to create tension and undermine local communities both internally and in their relationships with outsiders. Many practitioners call for caution. They note that the use of mapping tools and practice at the community level may lead to increased conflict, resource privatization, and loss of common property (Abbot et al., 1998; Crawhall, 2001; Fox et al., 2005; Harris and Weiner, 1998; McCall, 2004; Rundstrom, 1995).

For example, Fox et al. concluded after a two-year study of participatory mapping projects in Asia that 'spatial information technology transforms the discourse about land and resources, the meaning of geographical knowledge, the work practices of mapping and legal professionals and ultimately the very meaning of space itself' (2005: 10). The paper further argues that 'communities that do not have maps become disadvantaged as rights and power are increasingly framed in spatial terms' (2005: 7) and concludes on a critical note that mapping has become necessary – as failing to be on a map corresponds to a lack of proof of existence and to lack of ownership of land and resources. Overall, this must be framed in the need for developing 'critical clarity with respect to mapping based on a comprehensive understanding of both intended and likely unintended consequences of our actions' (2005: 10).

Thus, several decades of research and practice suggests that community mapping differs from conventional cartographic approaches in its processes, potential productions or outputs, and content – the sources and forms of spatial knowledge that are integrated. Some key statements used to recognize and denote community maps include the following.

Community mapping is defined by the process of production

Community maps are ideally planned around a consensus-based goal and strategy for use (Alcorn, 2000) and made with input from a whole community in an open and inclusive process (Aberley, 1993; Flavelle, 2002; Johnson, 1997). The higher the level of participation by all members of the community, the more beneficial the outcome, because the final map will reflect the collective experience of the group producing the map (Brody, 1981). This level of community engagement is of greater significance in the creation and effective use of community maps than the cartographic skills required to make them. Good practice associated with community mapping usually involves providing the spaces to enable everyone to take part in the map creation process, including women, youth, and the more powerless members of the community, without fear of having their views altered or manipulated by the more powerful within the community (Corbett and Keller, 2005; Flavelle, 2002; Rambaldi et al., 2006a; Sirait et al., 1994). While this ideal is not always met, the views of marginal groups might remain unheard if facilitators from outside the community do not create these spaces.

Community mapping is defined by a product that represents the agenda of the community

It is map production undertaken by communities to communicate information that is relevant and important to the community's needs and is for use by, or on behalf of, the community. Yet the challenge of community mapping processes is to ensure that most of the agendas of the community are included in the final maps and that these reflect the views of all relevant groups (both powerful and marginal). This can be challenging, particularly in the South. During community mapping work in East Kalimantan it became obvious that women's views of important spatial information and knowledge were focused on resources and features close to the village site, while men concentrated on boundaries and tenure-related issues on the periphery of traditional lands (Corbett, 2003). Presenting these disparate views and associated values on the same map can be challenging and can contribute to internal tensions, and occasionally conflict. Furthermore, there is the issue of whether a community mapping process involving an external agent or organization might be undertaken to further support and strengthen the agenda of the external agent to the detriment of the community.

Community mapping is defined by the content of the maps, which depict local knowledge and information

Community maps contain the community's place names, symbols, scales, and priority features (Flavelle, 2002; Orlove, 1993) and represent local knowledge systems via a locally defined visual language made explicit via the map legend (Rambaldi, 2005). Yet careful consideration needs to be given to how the content of the map might be used. Once a map has been created it is often put into the public domain. This turns local knowledge into public knowledge and conceivably takes it out of local control (Abbot et al., 1998). Documenting sensitive information and presenting it on a map might serve to make that information more vulnerable to exploitation; this is particularly the case when maps draw attention to high-value natural resources or archeological sites (Flavelle, 2002; Rambaldi et al., 2006a; Stockdale and Corbett, 1999). Maps make this information visible to outsiders, and therefore open to misuse. Furthermore, there might be information within the community that is 'owned' by certain individuals and families; this information cannot be shared with other community members, let alone decision makers and other groups from outside the community.

Community mapping is not defined by the level of compliance with formal cartographic conventions

Community maps are not confined by formal media; a community map may be incorporated into a GIS, a cardboard terrain model, or a drawing in the sand. Whereas maps made in the Western/Northern tradition seek conformity (Edney, 1993), community maps embrace diversity of presentation and content. Indeed idiosyncrasy and variety have been encouraged in some cases (Wright et al., 1997). Conversely a community map might lose its effectiveness to influence decision making if it is not presented in a format or through a medium that is considered formal or professional, risking the loss of legitimacy.

Thus, community mapping is process focused, with emphasis upon participant-determined goals, content, and representations. With these characteristics in mind, the following sections illustrate the multiple purposes or applications that tend to motivate community mapping, the diverse tools and techniques that are employed, and the complex outcomes that are often produced.

RANGE OF USES

Community mapping has been implemented in a broad spectrum of contexts. These include collaborative research initiatives (Hampson et al., 2003; Quan et al., 2001; Trong et al., 2002) community-based planning and monitoring (Bersalona and Zingapan, 2004; McCall, 2004; Rambaldi et al., 2002), asserting territorial claims, managing land-related disputes and supporting associated negotiations (Chacon, 2003; Cook et al., 2003; Wood, 2002), preserving and revitalizing indigenous cultural resources and intangible heritage (Crawhall, in press; Rambaldi et al., 2007), and undertaking consultative policy making (Carton, 2002a).

There are three main purposes in initiating a community mapping project. These ultimately relate to the need to communicate land-related knowledge: within communities, between neighboring communities, and from communities to outsider groups. These points are addressed in turn below.

Communicating information within communities

There are a number of ideal outcomes intended from a community mapping initiative. Perhaps one of the loftiest is for the mapping process to contribute to building community cohesion (Alcorn, 2000; Corbett and Keller, 2005; Stan and Amiel, in press) through providing a medium that allows a community to discuss and document its land-related knowledge. When elders share traditional place names and histories with other members of the community through the map-making process, it can generate a resurgence of interest in their local knowledge (Harmsworth, 1998) and facilitate intergenerational empathy (Corbett and Keller, 2004; Rambaldi et al., 2007). This can help a community sustain a sense of place and a connection to the land (Aberley, 1993; Chapin and Threlkeld, 2001).

The map-making process can also act as a focus for discussions that will assist with recognizing concerns and issues within the community. Discussions might raise community awareness about local and regional environmental issues or amplify community capacity to manage and protect lands (Bujang, 2005; de Vera, 2005; Hardcastle et al., 2004; Poole, 1995; Zingapan and de Vera, 1999). During the course of these discussions a community can formulate a common vision, which in turn may help to develop an effective community-based plan for future development (Harrington, 1995).

Community mapping is not about being an expert cartographer, but about community building, networking, and communication. Once a community has an articulated vision and representation of its identity within the context of its physical, biological, economic, and cultural landscapes, ideally it will be in a stronger position to effectively communicate and deal with external agencies, and it will be more likely to be involved in planning for its own future.

Community maps might also become a medium that allows communities to record and archive local knowledge. Local communities and indigenous groups are increasingly using community maps as a means to record, store, and manage important local knowledge and cultural information. Under threat from development and change, indigenous groups have used mapping projects to collect and preserve cultural histories

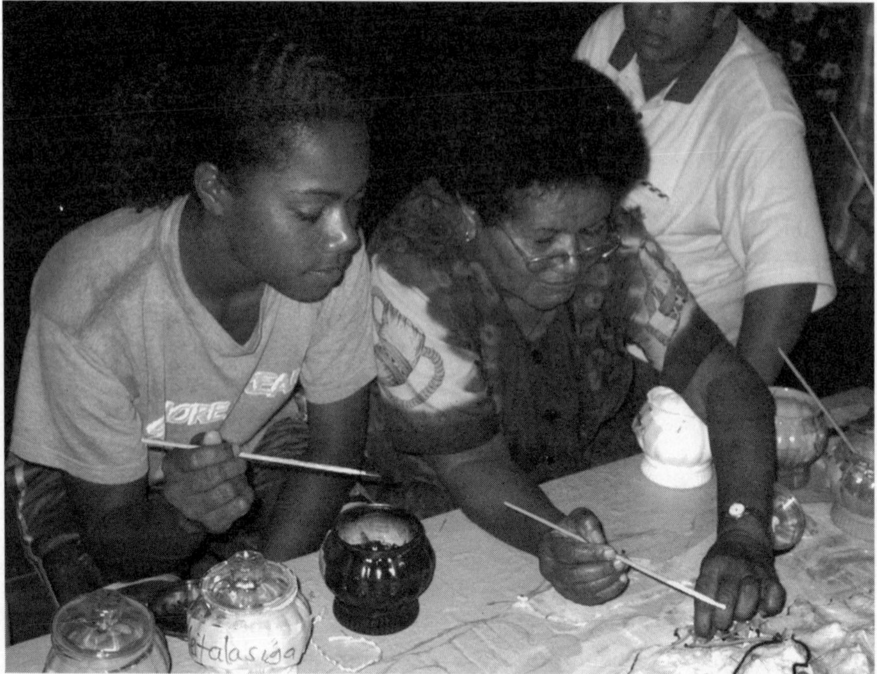

Figure 5.3 Fijian elders sharing knowledge with a student while marking resource areas on a 1:10,000 scale relief map of Ovalau Island, Levuka, Fiji Islands, April 2005

(Crawhall, in press; Harrington, 1995; Rambaldi et al., 2006c; 2007) and to record the knowledge of their elders about the land (Flavelle, 1995). This information is being recorded (Figure 5.3) for fear that it will otherwise be lost as the older generations pass away and traditional ways of life change.

However, the ideal outcomes noted above are by no means assured during a mapping process. Shared decision making and visions are sometimes irreconcilable even (and often especially) at the community level.[4] Furthermore, processes leading to ideal outcomes are often facilitated (and in turn the results are then shared) by 'experts' from outside the community, which again raises the issue of whose agenda is incorporated into the map, the map-making process and the communication of the results.

Communicating information between neighboring communities

Increasingly communities in the South are producing maps that are used to record land-related agreements and communicate information between communities. Uncertainty and flux relating to land rights, as well as increasing conflict over natural resources, have encouraged communities to map the extent of their traditional lands. In particular, this form of community mapping focuses on boundaries of negotiation and determination (Chapin and Threlkeld, 2001; Chapin et al., 2005).

For example, during the establishment of map-based community information systems in East Kalimantan in Indonesia, members from the participating communities of Benung and Tepulang decided it was necessary to document the location of the boundary between the two villages. On a prearranged day, elders from both villages met and walked the boundary between the two villages, agreeing on the position of the boundary without conflict. Using a global positioning system (GPS) and video camera, people from Tepulang recorded the entire process.

Recording spatial knowledge and community understandings of boundaries has the potential to allow conflicts to be solved more easily; however, it can also bring in problematic formalizations of geographic boundaries that create new layers of discord. Six months after the boundary map in East Kalimantan was made, Tepulang members began logging in the vicinity of the boundary between the two villages. Soon after operations had commenced, Benung claimed that the logging operation was straying onto their territory. A joint village meeting was called. The map, GPS, and video material captured during the boundary walk were displayed in this large meeting. Much discussion emerged related to the previously documented information, with all community members in the meeting constantly returning to the multimedia and map data for reference to support their arguments. As a result of the community meeting and the mapping material, the conflict was resolved; the logging operations withdrew from the contentious area, leaving the felled timber behind. However, several community members from Tepulang expressed a high level of frustration with the meeting's conclusion; they blamed the unfavorable outcome on the 'inaccurate' information contained on the community map and associated multimedia information, as well as the entire community mapping process within their community (Corbett and Keller, 2004). For the community members on the losing end of the 'map's ability to influence decision-making', it was an overwhelmingly negative and marginalizing process. Overall, however, the mapping process can help identify and document divisive issues so that they do not interfere with a larger consensus building process.

Communicating information to outsider groups

Community maps have proved to be an effective, powerful, and convincing medium to demonstrate to external agencies how a community values, understands, and interacts with its immediate space (Fox et al., 1996; Peluso, 1995; Poole, 1995; Rambaldi et al., 2006b). They have helped communities to communicate their often 'long but invisible history of managing resources' (Hitchcock, 1996; cited in Alcorn, 2000: 9). Provided community members have generated the map and the legend, community maps present complex information in a well-understood and easily accessible format. This enables groups with language and cultural barriers and differences in values to more easily communicate and understand the information presented. In the words of Aberley, 'maps can show a vision … more clearly than thousands of words' (1993: 4).

Communication of community spatial information can ultimately enable communities to apply pressure for change. In some cases maps have been used to request ownership over areas of customary land that have been claimed by the state (Bersalona and Zingapan, 2004; Bujang, 2005; Denniston, 1994; de Vera, 2005; Nietschmann, 1995). For example, the Gitxsan and Wet'suwet'en First Nation bands in British Colombia, Canada have used maps in their attempts to have their native sovereignty recognized by provincial and federal governments (Olive and Carruthers, 1997; Sparke, 1998). Community maps have become a tool with which communities can seek recognition and inclusion in governance and decision-making processes, particularly in reference to land and natural resource management (Fox, 1994; Sirait et al., 1994). At times they have also succeeded in empowering grassroots efforts to hold governments accountable. In this sense, map making is a form of political action that is capable of bringing about change.

RANGE OF TOOLS

A broad and growing number of community mapping tools are now available, and the choice of which one to use will ultimately be determined by the way in which the map will be employed and by the tool's ability to maximize the intended impact on the target audience. The tools chosen need to be appropriate for the available resources (financial, human, and equipment). The decision of which tools to use might also be influenced by whether the mapping initiative occurs within a community-driven process or whether it occurs within an externally driven process.

These tools range from low-cost, low-resource-input activities (such as sketch maps drawn in the sand – referred to as ephemeral mapping), through medium cost, medium resource input activities (such as the participatory three-dimensional modeling seen in Figure 5.4, or the production of a scale map using basic surveying techniques), to high-cost and high-resource-input programs (such as developing and deploying computer-based participatory geographic information systems). Although all these cartographic tools are now being used in community mapping initiatives, there exists an inverse relationship between the technology used and the levels of participation attained: the greater the cost and complexity of use of the mapping tool, the less likely community members will be to participate in the mapping initiative.

Although community maps have proven to be useful tools for communicating local knowledge, they are limited in describing the complexity and extent of what is known about the land. For this reason maps are frequently supplemented with the written word. This is often an imperfect medium to represent local knowledge, especially for traditional people who may be illiterate and accustomed to communicating orally. Johnson (1992) noted that much local knowledge about the land is transmitted in the form of stories and legends that use metaphor and sophisticated terminology that might be lost if the information is transcribed. In Northern Canada, Inuit groups believe that the written word fails to capture the depth and power of the spiritual relationship with the land (Johnson, 1992). There is a need for a tool that can combine the usefulness of maps with other digital media, such as video, images, and audio, which are better at documenting the oral and visual aspects and complexities of local knowledge.

Figure 5.4 Ogiek peoples visualizing their traditional lands using a physical 1:10,000 scale three-dimensional cardboard model, Nessuit, Kenya, 2006

Some practitioners argue that geographic information technologies (GIT), particularly GIS, can help demonstrate the close relationship between local people and their land by illustrating the multiple dimensions of human–land relations and, as a result, are well suited to preserve, revitalize, and disseminate local knowledge (Corbett and Keller, 2006; Harmsworth, 1998; Poole, 1995). These technologies maintain the capacity of the Cartesian map to organize and reference spatial information, and combine this with the capability of linking to attribute databases and other information in the form of digital images, audio and video. Increasingly GIS technologies are being utilized to address land-related issues with examples springing up around the global South (for examples see volume 54 of *Participatory Learning and Action,* special issue 'Mapping for Change: Practice, Technologies and Communication). Interestingly these applications have usually been adopted without the significant redesign of GIS. To an extent this reflects the flexible nature of GIS software, in particular its inherent ability to combine spatially referenced media (video, photographs, and text) and other cartographic data. The emergent field of qualitative GIS with its focus on multimedia and mixed methods provides a useful framework to better conceptualize the potential application of GIS technology for community representation.

However, as a rule, the more advanced the technologies employed, particularly in relation to computer-based mapping tools such as GIS and internet-based mapping, the greater the risk of a community failing to take ownership and long-term management of the maps and the tools and the processes being driven and controlled by external agents (see Elwood, this volume). Furthermore, the more technologically advanced the mapping system, the greater the long-term resources required (human, financial, and equipment) to update and maintain those mapping systems. This calls into question the long-term sustainability of these projects (Fox et al., 2003; Poole, 2006). However, this needs to be weighed against the possibility that the potential impact and persuasiveness of the map product might be stronger when presented in a digital medium than using less formal cartographic tools such as ephemeral and sketch mapping. Finding a balance between the intended purposes of the map, the available resources, the capacity in the community, and the duration of commitment to the project is vital to achieving a successful community mapping initiative.

RANGE OF IMPACTS

The ideal outcomes noted above are influenced in large part by a number of interacting factors including the presence of enabling or disabling environments, the role of technological intermediaries, and the complexity of managing relationships between the actors involved in the mapping process.

Presence of enabling or disabling environments

A formidable challenge to the realization of the potential offered by community mapping applications is the widespread lack of effective administrative mechanisms and structures by which decisions reached through community mapping and participatory GIS applications could be incorporated into mainstream decision making (Kyem, 2004).

Although in some countries legislation has created the space for community mapping practice to be operational and the map products to be fed directly into land use planning activities (e.g. Mozambique), lack of enabling environments or in some cases the presence of *disabling* legal and regulatory instruments (e.g. Malaysia) present a serious obstacle to the widespread adoption, application, and ultimately influence of community mapping products (Bujang, 2005).

For example, in the Philippines, the 1997 Indigenous Peoples Rights Act (IPRA) law established the rights of indigenous peoples to file claims and secure titles over ancestral lands or domains. The law institutionalized the leading role of the community by adopting the principle of 'self-delineation' in the conduct of all mapping and survey activities of traditional lands and territories. A year later this was challenged by the Philippine Geodetic Engineering Act of 1998, or Republic Act no. 8560, regulating mapping practice and limiting the use of geodetic instruments, the conduct of land surveys, and the preparation of geographic information systems to licensed geodetic engineers (de Vera, 2005), *de facto* signaling that the work of community mappers was outside the framework of the law. Accordingly, the disconnection between formal (government) and traditional (community) institutions may have to be reconciled first in order to facilitate enabling environments that allow for effective community mapping to take place (Rambaldi et al., 2006b).

There is a reciprocal relationship between community mapping and 'good governance'. An environment of good governance and the underlying (though elusive) value of 'political will' are necessary preconditions for community mapping to function in a meaningful and effective manner. However, concomitantly community mapping can also support effective good governance; it can be a practical mechanism that helps stimulate accountability, legitimacy, transparency, responsiveness, participation, respect for rights, equity, local usability, and other dimensions of good governance (McCall, 2004).

Roles and obligations of technology intermediaries

Producing georeferenced information from local spatial knowledge and rendering it in the form of authoritative maps depends on the availability of data capturing skills, equipment, and software. On top of that, converting resulting information into effective messages for advocacy or negotiation requires communication and networking skills that are above and beyond those skills required or generated during the community mapping process. It is evident that the skills needed to accompany a demand-driven and effective community mapping process are multidisciplinary and are possibly delivered by trained technology or advocacy intermediaries operating from within the community or working closely on their behalf. Considering the opportunities and threats resulting from documenting, georeferencing, and visualizing local knowledge, these intermediaries have to operate within the confines of a code of good practice (Rambaldi et al., 2006a). Although some informal codes of conduct do exist, there remains an enormous range of approaches and consequently ethical behavior in the application and delivery of community mapping initiatives.

Each profession and culture carries specific moral parameters and codes of ethics. As community mapping is a multidisciplinary practice it has to respond to a blend of different principles including, but not limited to, the following imperatives:

- selecting GIT which are adapted to local environmental conditions and human capacities
- obtaining prior informed consent
- putting local values, needs, and concerns first
- avoiding the raising of false expectations
- being considerate in taking people's time
- considering map making and maps as a means and not an end
- stimulating spatial learning and information generation rather than simply supporting data extraction for outsiders' analysis and interpretation (Rambaldi et al., 2006a).

CONCLUSIONS

Community mapping initiatives in the global South continue to expand in both extent and scope. Furthermore, local communities and their partner organizations appear committed to innovate and use new and emergent mapping tools and technologies to succeed in their desired aims of representing, documenting, and communicating community knowledge in order to influence land-related decision making. This involves using a range of tools, from ephemeral mapping to developing and deploying sophisticated qualitative GIS systems that combine mixed multimedia material (including video and image data) volunteered from multiple sources (men, women, young and old) with digital maps.

Despite the technology, what ultimately defines community mapping in the South is the processes by which the maps are made and put to work. Almost without exception, these initiatives require the development of linkages with groups outside the community to assist in the map-making and follow-up activities. As a result, community mapping initiatives become as (if not more) contingent upon networking as upon cartographic expertise. In turn, this means that mapping processes are often facilitated by outsider groups (including non-governmental organizations, community-based organizations, universities, and development agencies) who are ideally strongly committed to the principles of participatory development and high levels of local community engagement. Ideally, too, they consciously maintain the position of 'facilitator' and not that of the 'expert', which requires skill and delicacy.

Community mapping initiatives are complex processes and rife with contradictions; they have the ability to heighten the tensions and uncertainty that they seek to resolve. This is not because they are inherently flawed, but rather because they reflect the complex roles and associations that already exist in every community throughout the world. However, given the uncertainty of land rights and ownership in the South, community mapping remains one of the most important practices in enabling local communities to represent their own relationship to their territories, and subsequently to engage in and influence land-related decision-making processes. Community mapping's power and contribution to social justice should, therefore, not be underestimated.

NOTES

1 The products of the exercise (video clips, images, audio recordings, notes, written statements, drawings, diagrams drawn by the villagers, and other forms of data) plus data gathered via other exercises are ultimately intended to be compiled in a multimedia Ogiek atlas.

2 Within the context of this chapter, 'local community' is defined as a group of people who regularly associate with one another in a specific geographic location on the basis of shared interest, reliance, relations, and identity. When using this definition the authors recognize that the 'community' is not a homogeneous entity but rather an affiliation of individuals, and that 'communities are differentiated in terms of status, income and power' (Midgley, 1986: 35).

3 For community mapping case examples from around the world, see the following resources: Southeast Asia, Bujang (2005), Flavelle (1995), Momberg et al. (1994), Peluso (1995), and de Vera (2005); Central Asia, Jackson et al. (1994), Poffenberger (1996); Africa, Harris and Weiner (1998), Jackson and Bond (1997), and Rambaldi et al. (2007); Europe, Carton (2002a; 2002b) and King (1993); North, South, and Central America, Bird (1995), Chapin and Threlkeld (2001), Jardinet (2006), Kemp and Brooke (1995), and Poole (2006); the Pacific, Rambaldi et al. (2006c); and Australasia, Harmsworth (1998).

REFERENCES

Abbot, J., Chambers, R., Dunn, C., Harris, T., De Merode, E., Porter, G., Townsend, J. and Weiner, D. (1998) 'Participatory GIS: opportunity or oxymoron?', *PLA Notes*, 33: 27–33.

Aberley, D. (1993) *Boundaries of Home: Mapping for Local Empowerment*. Gabriola Island: New Society.

Alcorn, J.B. (2000) 'Keys to unleash mapping's good magic', *PLA Notes*, 39: 10–13.

Belyea, B. (1992) 'Images of power', *Cartographica*, 29: 1–9.

Bersalona, R. and Zingapan, K. (2004) 'P3DM: mapping out the future of indigenous peoples in 3D', *ICT Update*, 17 (May). http://ictupdate.cta.int/index.php/article/articleview/305/1/56/.

Bird, B. (1995) 'The EAGLE project: re-mapping Canada from an indigenous perspective', *Cultural Survival Quarterly*, 18: 23–4.

Brody, H. (1981) *Maps and Dreams: Indians and the British Columbia Frontier*. Toronto: Douglas and McIntyre.

Bujang, M. (2005) 'A community initiative: mapping Dayak's customary lands in Sarawak, Malaysia', paper presented at the Mapping for Change Conference, Nairobi, Kenya.

Carton, L. (2002a) 'Strengths and weaknesses of spatial language: mapping activities as debating instrument in a spatial planning process', paper presented at the FIG XXII International Congress, Washington, DC.

Carton, L. (2002b) 'The power of maps in interactive policymaking: visualization as debating instrument in strategic spatial policy processes', in E.F.T. Heuvelhof (ed.), *The First International Doctoral Consortium on Technology, Policy and Management*. Delft, Netherlands: TU Delft.

Chacon, M. (2003) 'Principles of PPGIS for land conflict resolution in Guatemala', paper presented at the Geography Department, UCGIS Summer Assembly, Pacific Grove, California, 16–20 June.

Chapin, M. and Threlkeld, B. (2001) *Indigenous Landscapes: A Study in Ethnocartography*. Arlington, VA: Center for the Support of Native Lands.

Chapin, M., Lamb, M. and Threlkeld, B. (2005) 'Mapping indigenous land', *Annual Review of Anthropology*, 34: 619–38.

Cook, S., O'Brien, R., Corner, R. and Oberthur, T. (2003) 'Is precision agriculture irrelevant to developing countries?', paper presented at the European Conference on Precision Agriculture, Berlin, Germany.

Corbett, J. (2003) 'Empowering technologies? Introducing participatory geographic information and multimedia systems in two Indonesian communities', paper presented at the Department of Geography, University of Victoria, Victoria, Canada.

Corbett, J.M. and Keller, C.P. (2004) 'Empowerment and participatory geographic information and multimedia systems: observations from two communities in Indonesia', *Information Technologies and International Development*, 2: 25–44.

Corbett, J.M. and Keller, C.P. (2005) 'An analytical framework to examine empowerment associated with participatory geographic information systems (PGIS)', *Cartographica*, 40: 91–102.

Corbett, J.M. and Keller, C.P. (2006) 'Using community information systems to express traditional knowledge embedded in the landscape', *Participatory Learning and Action*, 54: 21–7.

Crampton, J. (1995) 'The ethics of GIS', *Cartography and Geographic Information Systems*, 22: 84–9.

Crawhall, N. (2001) 'Written in the sand: cultural resources auditing and management with displaced indigenous people'. Unpublished manuscript. Cape Town, South Africa, The South African San Institute (SASI) in cooperation with UNESCO.

Crawhall, N. (in press) *Cultural Mapping and Indigenous Peoples: Promoting Intercultural Dialogue and Cultural Diversity*. Paris, France: UNESCO.

Dahl, E.D. (1992) 'Commentary: Brian Harley's influence on modern cartography', *Cartographica*, 29: 62–5.

Denniston, D. (1994) 'Defending the land with maps', *World Watch*, 7: 27–31.

De Vera, D. (2005) 'Mapping with communities in the Philippines: rolling with the punches', paper presented at the Mapping for Change Conference, Nairobi, Kenya.

Edney, M. (1993) 'The patronage of science and the creation of imperial space: the British mapping of India, 1799–1843', *Cartographica*, 30: 61–7.

Flavelle, A. (1995) 'Community-based mapping in Southeast Asia', *Cultural Survival Quarterly*, 18: 72–3.

Flavelle, A. (2002) *Mapping Our Land: A Guide to Making Maps of Our Own Communities and Traditional Lands*. Edmonton, Canada: Lone Pine Foundation.

Fox, J. (1994) *Spatial Information and Ethnoecology: Case Studies from Indonesia, Nepal, and Thailand*. Honolulu: East–West Center.

Fox, J., Yonzon, P. and Podger, N. (1996) 'Mapping conflicts between biodiversity and human needs in Langtang National Park, Nepal', *Conservation Biology*, 10: 562–9.

Fox, J., Suryanata, P.K. and Pramono, H. (2003) 'Mapping power: ironic effects of spatial information technology', in *Spatial Information, Technology and Society: Ethics, Values, and Practice Papers*. Honolulu: East–West Center.

Hampson, K., Bennett, D., Alviola, P., Clements, T., Galley, C., Hilario, M., Ledesma, M., Manuba, M., Pulumbarit, A., Reyes, M.A., Rico, E. and Walker, S. (2003) *Wildlife and Conservation in the Polillo Islands, Philippines. Polillo Project Final Report.* Glossop, UK: Viper. http://mampam.50megs.com/polillo/2001/index.htm.

Hardcastle, J., Rambaldi, G., Long, B., Lanh, L.V. and Quoc, S.D. (2004) 'The use of participatory 3-dimensional modeling in community-based planning in Quang Nam province, Vietnam', *Participatory Learning and Action*, 49: 70–6.

Harley, J.B. (1988) 'Maps, knowledge and power', in D. Cosgrove (ed.), *The Iconography of Landscape.* Cambridge, MA: Cambridge University Press.

Harley, J.B. (1989) 'Deconstructing the map', *Cartographica*, 26: 1–20.

Harley, J.B. (1990) 'Cartography, ethics and social theory', *Cartographica*, 27: 1–23.

Harmsworth, G. (1998) 'Indigenous values and GIS: a method and a framework', *Indigenous Knowledge and Development Monitor*, 6 (3): 1–7.

Harrington, S. (1995) *Giving the Land a Voice: Mapping Our Home Places.* Salt Spring Island, BC: Salt Spring Island Community Services Society.

Harris, T. and Weiner, D. (1998) 'Empowerment, marginalization, and "community-integrated" GIS', *Cartography and Geographic Information Systems*, 25: 67–76.

Harvey, D. (1990) *The Condition of Postmodernity: An Inquiry into the Origins of Cultural Change.* Cambridge: Blackwell.

Hitchcock, R. (1996) 'Kalahari communities: Bushmen and the politics of environment in Southern Africa'. International Working Group for Indigenous Affairs (IWGIA), Copenhagen.

Jackson, B. and Bond, D. (1997) 'Monitoring and evaluating collaborative management of natural resources in Eastern Africa'. Workshop Report, Tanga Coastal Zone Conservation and Development Programme, Tanga, Tanzania, 7–11 April.

Jackson, B., Nurse, M. and Singh, H.B. (1994) 'Participatory mapping for community forestry'. Rural Development Forestry Network, Network Paper, 17e, pp. 1–11.

Jardinet, S. (2006) 'Capacity development and PGIS for land demarcation: innovations from Nicaragua', *Participatory Learning and Action,* 54: 67–73.

Johnson, B. (1997) 'The use of geographic information systems (GIS) by First Nations'. School of Community and Regional Planning, University of British Columbia.

Johnson, M. (1992) *Lore: Capturing Traditional Environmental Knowledge.* Ottawa, ON: International Development Research Centre.

Kemp, W.B. and Brooke, L.F. (1995) 'Towards information self-sufficiency: Nunavik Inuit gather information on ecology and land use', *Cultural Survival Quarterly*, 18: 25–8.

King, A. (1993) 'Mapping your roots: parish mapping', in D. Aberley (ed.), *Boundaries of Home.* Gabriola Island: New Society.

Kyem, P.A.K. (2004) 'Power, participation, and inflexible institutions: an examination of the challenges to community empowerment in participatory GIS applications', *Cartographica,* 38: 5–18.

Livingstone, D. (1992) *The Geographical Tradition*. Oxford: Blackwell.

McCall, M. (2004) 'Can participatory GIS strengthen local-level planning? Suggestions for better practice', paper presented at the 7th International Conference on GIS for Developing Countries (GISDECO 2004), Universiti Teknologi Malaysia, Johor Malaysia.

Midgley, J. (1986) 'Community participation: history, concepts and controversies', in J. Midgley, A. Hall, M. Hardiman and D. Narine (eds), *Community Participation, Social Development and the State*. London: Methuen. pp. 13–16.

Momberg, F., Damus, D., Limberg, G. and Padan, S. (1994) *Participatory Tools for Community-Forest Profiling and Zonation of Conservation Areas: Experiences from the Kayan Mentarang Nature Reserve, East Kalimantan, Indonesia.* Washington, DC: WWF Indonesia Program.

Monmonier, M. (1991) *How to Lie with Maps.* Chicago: University of Chicago Press.

Nietschmann, B. (1995) 'Defending the Miskito reefs with maps and GPS', *Cultural Survival Quarterly*, 18: 34–6.

Olive, C. and Carruthers, D. (1997) *Putting TEK into Action: Mapping the Transition*. Vancouver, BC: Ecotrust Canada Mapping Office.

Orlove, B. (1993) 'The ethnography of maps: the cultural and social contexts of cartographic representation in Peru', *Cartographica,* 30: 29–45.

Peluso, N.L. (1995) 'Whose woods are these? Counter-mapping forest territories in Kalimantan, Indonesia', *Antipode*, 27: 383–406.

Poffenberger, M. (1996) *Grassroots Forest Protection: Eastern Indian Experiences*. Bohol, Philippines: Asia Forest Network.

Poole, P. (1995) 'Geomatics, who needs it?', *Cultural Survival Quarterly*, 18 (4). http://209.200.101.189.offcampus.lib.washington.edu/publications/csq/csq-article.cfm?id=1242.

Poole, P. (2006) 'Is there life after tenure mapping?', *Participatory Learning and Action*, 54: 41–59.

Quan, J., Oudwater, N., Pender, J. and Martin, A. (2001) 'GIS and participatory approaches in natural resources research', in *Socio-economic Methodologies for Natural Resources Research Best Practice Guidelines*. Chatham, UK: Natural Resources Institute.

Rambaldi, G. (2005) 'Who owns the map legend?', *URISA Journal*, 17: 5–13.

Rambaldi, G., Bugna, S., Tiangco, A. and deVera, D. (2002) 'Bringing the vertical dimension to the negotiating table: preliminary assessment of a conflict resolution case in the Philippines', *ASEAN Biodiversity*, 2: 17–26.

Rambaldi, G., Chambers, R., McCall, M. and Fox, J. (2006a) 'Practical ethics for PGIS practitioners, facilitators, technology intermediaries and researchers', *Participatory Learning and Action*, 54: 106–13.

Rambaldi, G., Kyem, P.K., Mbile, P., McCall, M. and Weiner, D. (2006b) 'Participatory spatial information management and communication in developing countries', *EJISDC*, 25: 1–9.

Rambaldi, G., Tuivanuavou, S., Namata, P., Vanualailai, P., Rupeni, S. and Rupeni, E. (2006c) 'Resource use, development planning, and safeguarding intangible cultural heritage: lessons from Fiji Islands', *Participatory Learning and Action*, 54: 28–35.

Rambaldi, G., Muchemi, J., Crawhall, N. and Monaci, L. (2007) 'Through the eyes of hunter-gatherers: participatory 3D modeling among Ogiek indigenous peoples in Kenya', *Information Development*, 23: 113–28.

Rundstrom, R.A. (1995) 'GIS, indigenous peoples, and epistemological diversity', *Cartography and Geographic Information Systems*, 22: 45–57.

Sirait, M., Prasodjo, S., Podger, N., Flavelle, A. and Fox, J. (1994) 'Mapping customary land in East Kalimantan, Indonesia: a tool for forest management', in J. Fox (ed.), *Spatial Information and Ethnoecology: Case Studies from Indonesia, Nepal, and Thailand*. Honolulu: East–West Center.

Sparke, M. (1998) 'A map that roared and an original atlas: Canada, cartography, and the narration of nation', *Annals of the Association of American Geographers,* 88: 463–95.

Stan, A. and Amiel, S. (in press) 'Rapport sur l'atelier: la cartographie culturelle et ses possibles applications par les peuples autochtones' paper presented at the UNESCO Workshop on Cultural Mapping, Paris, France.

Stockdale, M.C. and Corbett, J.M. (1999) *Participatory Inventory: A Field Manual Written with Special Reference to Indonesia*. Oxford: Oxford University Press.

Trong, H.T., Castella, J.C. and Eguienta, Y. (2002) 'Participatory 3-D landscape modeling: towards a common spatial language among researchers and local stakeholders', in T. Paopa (ed.)*, Scaling-Up Innovative Approaches in Agricultural Development*. Hanoi, Vietnam: Agricultural Publishing House.

Wood, B. (2002) 'GIS as a tool for territorial negotiations', *IBRU Boundary and Security Bulletin*, 8: 72–8.

Wood, D. (1992) *The Power of Maps*. New York: Guilford.

Wright, D.J., Goodchild, M.F. and Proctor, J.D. (1997) 'GIS: tool or science? de-mystifying the persistent ambiguity of GIS as "Tool" versus "Science"', *Annals of the Association of American Geographers*, 87: 346–62.

Wright, J.K. (1942) 'Map makers are human: comments on the subjective in maps', *Geographical Review*, 3: 527–44.

Zingapan, K. and de Vera, D. (1999) 'Mapping the ancestral lands and waters of the Calamian Tagbanwa of Coron, Northern Palawan', paper presented at the Conference on NGO Best Practice, Philippines.

ANALYTICAL INTERVENTIONS AND INNOVATIONS

6

GROUNDED VISUALIZATION AND SCALE: A RECURSIVE ANALYSIS OF COMMUNITY SPACES

LaDona Knigge and Meghan Cope

ABSTRACT

Grounded visualization is an approach to data exploration and analysis that enables researchers to examine spatially referenced data and ethnographic data together, in close relationship to one another, and to integrate GIS-based cartographic representations with qualitative forms of evidence and analysis. Prior discussion of grounded visualization has illustrated how this technique involves iterative reflexive engagement with different forms of data, enabling critical exploration of tensions and mismatches in different data and interpretations of them, to build stronger explanations. In this chapter, we extend earlier work on grounded visualization to illustrate how it can also support a multi-faceted engagement with scale as *both* cartographic representation *and* socio-political construction. Grounded visualization can support inductive, critically reflexive scale-sensitive analysis that will build richer and more nuanced understandings of, for example, contested urban spaces. We develop these arguments with examples from research in Buffalo, New York, in which we examine the constitution and contestation of 'vacant' urban land at multiple scales of analysis and representation, and through multi-scalar social, political, and economic processes.

INTRODUCTION

The distinct advantage of mixed methods in research, as developed in this volume and elsewhere (see Elwood, 2009), lies partly in the purposeful use of different methods to get at different aspects or questions in the course of a research project, but also in the productive tensions that can arise in allowing data from different methods to rub up against each other, conflict, complement, or even raise new questions. In mixed methods research, then, it is often helpful to have a sort of 'glue' to bind the methods together, an approach for bringing different methods or forms of evidence into productive interaction in research. Visualization offers one such possibility. Visualization is an approach in which data are explored visually using different techniques, including statistical analysis, charts and graphs, or – most relevant here – mapping exercises. This chapter builds on previous work (Knigge and Cope, 2006) to explore ways that ethnographic methods of qualitative research can be combined with visualization of spatial data to create new forms of knowledge based on the intentional, reflexive mixing of these approaches.

We begin with a review of the approach we call 'grounded visualization', then expand this discussion to consider how grounded visualization might encompass the inherently geographic issue of scale. We suggest that grounded visualization allows researchers to examine multiple conceptualizations of scale in the context of a single research effort: the scales of data, the production of scale through social action, and the use of cartographic scale in visualizing information. An empirical case study based on Knigge's research of vacant lots and community gardens in Buffalo, New York is then presented to instantiate the conceptual developments of grounded visualization and scale. In particular, we work through ways in which the seemingly binary land use category of 'vacant' in fact is striated in multiple ways, both in 'official' data[1] and as evidenced through local residents' everyday lives and interactions with these spaces. We suggest that the subtleties and power of these striations would have been much less apparent without the iterative, reflexive strategies of grounded visualization.

GROUNDED VISUALIZATION AND SCALE

Grounded visualization draws upon two analytical methods – grounded theory and visualization. Grounded theory is an approach emerging from sociology and anthropology that builds theory through multiple iterations of data collection, coding, categorization, comparison, and analysis to construct knowledge that is thoroughly 'grounded' in empirical data. Typical data production methods used in a grounded theory qualitative research project include talking with people (e.g. in focus groups and interviews), researchers' field notes, subjects' travel diaries, photographs, participatory mapping, and other media for eliciting and observing social information. However, there is no preclusion from using quantitative data in a grounded theory approach, so the use of spatially referenced numerical and categorical data is well suited to the broad-reaching, inclusive, analytical stance of grounded theory. Visualization generally includes the broad array of methods, mentioned above, that are used to visually explore and represent quantitative and qualitative data (usually 'official' data) using GIS or other software. For example, one visualization approach, exploratory spatial data analysis (ESDA), employs tools such as identifying trends on a map, displaying spatial autocorrelation, finding spatial outliers, and *smoothing*: 'Where the map consists of many small areas it is often helpful to apply simple smoothing methods which, depending on the scale of the smoother, may help to reveal the presence of general patterns that are unclear from the mosaic of values' (Haining and Wise, 1997). Mainstream GIS can also be used to perform ESDA operations such as focusing or brushing to highlight a portion of the dataset; linking maps with tables, graphs, or other forms of display; and multivariate mapping; as well as through various cartographic techniques such as changing category breaks, scale, color, and shading, and other tools to 'see' new patterns, which can in turn be explored further (Slocum et al., 2009).

Thus, *grounded visualization* builds upon and integrates the strengths and potentials of these two exploratory methods, enabling multiple queries of the data through an iterative, recursive process that blends qualitative and quantitative research, in this case from a geographic perspective. For the purposes of this chapter, we are especially interested in blending spatially referenced quantitative and categorical data with qualitative data about places and the uses and constructions of space.

To be more specific, in Knigge and Cope (2006) we identified four key areas of commonality between grounded theory and visualization of spatial data: both are exploratory, both are iterative and recursive, both incorporate the particular *and* the general, and both have the capacity to accommodate multiple interpretations of the world. In this chapter, we seek to expand on this common ground by building on the third area – 'simultaneous attention paid to both the particular and the general, the concrete and the abstract, and *the small and large scale*' (2006: 2028, emphasis added). Grounded theory has perhaps had too few encounters with scale-sensitive approaches, a gap that geographers are well equipped to rectify. Generally, the principles and practice of grounded theory analysis (particularly its 'constructivist' stream: see Charmaz, 2000) suit a critical engagement with the question of scale. This is similar to approaches in critical social geography, such as the understanding that social realities and meanings are perpetually socially constructed, lived within, and/or resisted. On the other hand, visualization techniques, in GIS and other spatially referenced media, typically employ a more cartographic sense of scale through such techniques as changing the level of resolution from coarse to fine. Geographers have identified many of the intersecting ways social realities are produced at multiple scales, and how places, regions, and locales (of various scales) are produced and imbued with meanings. Specifically, geographers' attention to the ways that social, economic, political, and environmental processes both construct and reify scales is helpful to begin understanding how grounded visualization can be of use.

In attempting to combine grounded theory and visualization techniques for geographic analysis with scale in the forefront, several dimensions emerge as relevant. First, there is the issue of the scale of the *data*, for example, census data collected at the scale of the household but aggregated to the block group or census tract level. Second, there is the scale of the *processes* that are of interest, particularly relevant for the ways that processes involve multiple scales. For instance, residential segregation by race represents a set of processes based on long-term and large-scale racial ideologies, but also based on (and perpetuated by) small-scale, micro-geographies of discriminatory practices by individuals, agencies, and municipalities. Third, in a more technical sense and somewhat in tension with the social constructivist views of the production of scale through processes, there is the scale of cartographic *representation*. A grounded visualization approach might engage this notion of scale through iterative exploration of maps at different scales, using the 'zoom' function available in cartographic software to see how the visibility or seeming significance of particular spatial patterns, clusters, or anomalies might change at different cartographic scales.

The issue of scale offers several fruitful opportunities to demonstrate the potential contributions of grounded visualization analyses, and illustrates how grounded visualization is *both* qualitative *and* quantitative, reflexively mixed together. The goal here is to demonstrate one example of a research study which, by paying attention to these three dimensions of scale in the analysis, allows the exploration of connections not immediately apparent, while also identifying the tensions between different knowledges as constructed through different research approaches.

The dimension of data: multiple realities

Spatially referenced data of all kinds now saturate our everyday lives, and new visualizations and uses for them in social research are emerging quickly (see Jung, this volume). While much of this information is quantitative, such as demographic and economic data, it often also has qualitative dimensions (see Pavlovskaya, this volume). For example, the complexities of how people identify their race and ethnicity on US census forms are well documented (Brewer and Suchan, 2001), but these are not apparent in maps that show *percentages* (a quantitative measure of proportion) of residents by race and ethnicity. 'Official' data are collected by government agencies and other research entities at a variety of *scales*, some of which are explicit (such as households, metropolitan areas, or state/province levels), and others of which are less obvious (such as data on trade flows, political opinions, or school performance). Additionally, qualitative data are collected (or 'produced') at various scales, many of which have been critically examined in recent geographic literature, such as the body, the household, the neighborhood, the region, the nation, or even the 'global' (McDowell and Sharp, 1997). Scale continues to matter beyond just the processes of data collection, into analysis and representation as well, meaning that attention to scale at all stages of research and reporting is perquisite to responsible scholarship.

A grounded visualization approach is valuable in addressing these concerns because of its attention to the diverse sources, types, and scales of data; such an approach involves thinking critically about the construction of the data, questioning the formation and processes of categorization, reflexively examining multiple datasets and how they relate, and using visualization techniques to explore consistencies and divergences. For example, the scale at which data are aggregated can produce a variety of problems with respect to appropriate inferences or conclusions that might be drawn from them. While the scale of the county gives us one bit of information about the material reality of a city, and data at the scale of the census tract give us another bit, we must be wary of generalizing, mixing, or inference. First, making the assumption that the attributes and relationships found at one level of aggregation (e.g. the census tract) can be applied to a more detailed level of aggregation (e.g. the household) or to individuals can result in what is referred to as the ecological fallacy. Thus, while the percentage of African Americans and the percentage of people living in poverty may be high within a census tract, we cannot infer that if an individual is African American, he/she must also be living in poverty. Second, the combination of spatial aggregation units into larger units can also result in significant errors in data, a problem referred to as the 'modifiable areal unit problem'. Further, data analysis and representation through visualization always involve a process of reduction, simplification, and smoothing in order to make sense of the world. As one team said, mapping "reduces the complexity of the real world into shaded area-patterns" (Fiedler et al., 2005: 149), but these reductions of complexity and their implications for data analysis and interpretation must also be scrutinized.

Critical geographers also question the way that census data are collected, arguing that the aggregation units, the scale, and the categories are socially constructed and politically influenced (Brown and Knopp, 2006; McMaster et al., 1997). When mapping census data, for instance, critical perspectives suggest staying aware of the power imbued by maps and questioning whether these representations conceal, subjugate, and/or silence certain features while foregrounding others (Harley, 1989).

Finally, other scale issues intrinsic in spatial data also arise when we wish to acknowledge and incorporate into our analysis the *multiple realities* of a place, an important goal of grounded visualization. The 'official' account of reality (as presented in 'official' sources of data) may not match the 'lived' reality. A key reason for these mismatches may be the different scales at which these data are produced, aggregated, and presented. For example, many urban scholars have noted the mismatch between census tracts and the localized, functional neighborhoods that are perceived and represented by residents (Sheppard, 1990), yet the practice of using census tracts as a proxy for neighborhood analysis persists. This is further complicated by subsequent research over time: the population of a neighborhood is aggregated along census block or tract designations that may change from decade to decade, with little regard for differently shifting subjective boundaries or the everyday lived experience of individual community members (Fiedler et al., 2005).

The social production of scale: questioning categories

Using visualization techniques to explore spatial data reveals patterns, clusters, and relationships, but qualitative methods help to understand how and why those patterns, clusters, and relationships exist. Grounded visualization's practice of questioning assumed categories and complicating taken-for-granted binaries allows researchers to consider a range of possible explanations for their findings. This also encourages scholars to engage with the *social production of scale* by identifying diverse realities at scales that are constructed by the intersecting operations of economic, social, and political processes. Scale is socially produced through both *material* processes (economic exchanges, social relations of culture and institutions, political processes of power), and *discursive* practices (geographic imaginaries of 'the neighborhood', 'the city', or 'the nation'), though these often conflict and generate tensions.

Feminist theory has been particularly helpful in this regard because of its tradition of challenging taken-for-granted assumptions and breaking down binaries and discrete categories, classifications, definitions, and concepts, such as men/women, public/private, rational/emotional (Women and Geography Study Group, 1997). Further, feminist geographers attempt to reveal unseen oppressions by challenging, problematizing, and contesting the ways that spatial divisions (such as home and work, or inside and outside) are central to the social construction of gender and power divisions (McDowell, 1999). And methodologically, feminists have been instrumental in developing new understandings of what constitutes geographic and social knowledge, as well as legitimating the perspectives of oppressed 'Others' through a range of research practices and awareness of mixed subjectivities (McLafferty, 2002; Moss, 2003).

Thus, scrutinizing the social production of scale involves critical analysis of multiple relations and processes, as well as an awareness of the ways scale is produced contingent upon *context*. As Sallie Marston notes, 'Theorization about scale attempts to understand how it is constructed or transformed within a particular historical and geographical context, and what implications particular manifestations of scale might have for social, cultural, political, and economic practices' (2004: 170). Feminist geographers have been instrumental in opening up new scales of analysis such as the household and the body, which highlight why context is so important: the meanings and constitution of what is a 'household' (or even a 'body') are not constant over time and place or across cultures. These advances could not be possible without thoroughly questioning assumed categories and binaries. This chapter uses a different type of taken-for-granted category (the land use classification 'vacant') as an example of how a critical scale-sensitive perspective rooted in grounded visualization can produce new understandings of geographic meanings.

The representation dimension: reflexive, iterative explorations

Grounded visualization also allows for critical reflexive engagement with scale by way of the cartographic representations of spatial data that can be produced with a GIS. One of the hallmarks of critically engaged visualization is its exploration of multiple cartographic representations to interrogate the similarities, differences, patterns, meanings, or gaps and omissions advanced in them. An examination of two differently scaled maps may well highlight important limitations of the data, provide the researcher with clues about the robustness of competing explanations, or suggest different interpretations of the phenomena being studied. Most significantly for *grounded* visualization, this interrogation also suggests new lines of inquiry in other realms, particularly through qualitative techniques for understanding the social processes that underlie the spatial patterns seen in cartographic visualizations. That is, in this kind of approach, visualization is always taken to be *one of many* research operations that reflect upon and inform each other in iterative rounds of data collection and interpretation. For example, we show how examining spatial patterns of urban land use and vacant lots spurred ethnographic research investigating why clusters of vacancy existed in particular neighborhoods and how local residents perceived and used these spaces. In this way, grounded visualization is not just a method of analysis, but also the means to connect spatially referenced data (in this case demographic and parcel data) with qualitative data generated through ethnography.

Geographic information systems are critical to enabling these sorts of investigations, because the technology makes it feasible for a researcher to carry out these explorations. Techniques include altering the spatial scale and resolution of the map; administering generalization processes including data aggregation, simplification, smoothing, amalgamation (fusing nearby polygons), and merging; and manipulating various aspects of how the data are represented in the map, such as changing data classification intervals, category breaks, or categorization methods on a choropleth map (Slocum et al., 2009). These features support the researcher's ability to play these different representations off one another in an inductive, *recursive* analysis process, to begin to draw out new types of synthesized knowledge.

Thus, in grounded visualization, critical engagements with cartographic scale are part of a mixed methods approach to building knowledge, such that data that may be considered largely quantitative can be linked with the processes that generate them. Simultaneously, qualitative data – such as people's experiences, transactions, relations, and actions – can be substantively linked with spatial patterns to provide mutually reflexive insight into effects of local and regional context, outcomes of societal processes, impacts of policy decisions on lived realities, and implications of discourse and ideology on people's experience of place.

This type of inductive, scale-sensitive analysis that can be fostered by grounded visualization is part of the technique's critically reflexive and intentional engagement with moments of fixity in data representation and analysis. As discussed earlier, certain aspects of the 'fixity' of cartographic scale in maps and spatial data present challenges for interpretation and analysis: the modifiable areal unit problem, spatial autocorrelation, the ecological fallacy, and so on. Without denying these challenges, grounded visualization is committed to productive engagement with these moments of fixity, while also exploring the fluidity of material and discursive experiences.

QUESTIONING VACANCY: A GROUNDED VISUALIZATION CASE STUDY

The theme of questioning binaries or discrete categories is exemplified here using the classification of 'vacancy' for land parcels, particularly those in urban settings where past, present, and potential future uses are highly dependent on local context. 'Vacancy' is a term ripe for unpacking because its seeming binary characteristic (vacant or not vacant?) masks a host of different – and often conflicting – meanings and uses of real space. Part of this difference in the meanings of 'vacant' lies in the clash between official data and lived experiences, but as we document, even the official classification of 'vacant' is inherently striated with different meanings, potential futures, and objective attributes. We demonstrate here that a scale-sensitive perspective within mixed methods grounded visualization enables a more nuanced and revelatory analysis of contested urban spaces.

The construction of state sponsored or 'official' knowledge occurs in part through the assigning of attributes and the categorization of data into discrete categories. Thus, the designation of a parcel of land as *vacant* is central to the collection of state sponsored 'official' data. However, depending upon the context, the term has multiple definitions. For example in Buffalo, New York, according to one report, 'vacant' may represent a binary distinction referring to occupancy that is 'linked with the property *use* or most recent past use or condition, i.e. vacant house, vacant residential lot, abandoned railway, former industrial site, brownfield' (Cornell Cooperative Extension, 2004: 23, emphasis added). The report, which lists 42 different vacant property types, also states that the term may refer to open, undevelopable space reserved for recreation, environmental functions, or natural uses as in the case of riparian, shoreline, wetlands, or floodplains. At a different scale of governance, the term *vacant* may represent one of the nine official discrete land use categories regulated by New York State which include agricultural, residential, commercial, community services, industrial, and other uses, designated by a numeric code that is attached to the parcel description and tracked by the City through property tax assessment rolls (New York State Office of Real Property Services, 2006). This type of vacant land, void of improvements, is inventoried and compiled by the City of Buffalo into a database of 'shovel-ready' developable sites. Discursively, these appear in reports and policy documents based on interim uses that can 'hold the land in manageable condition', which in various ways convey benefits to the immediate community and the City of Buffalo.

While on one level, the designation of 'vacant' has a set of discrete, categorical, land use *definitions*, further exploration reveals many varied *meanings* of vacancy, depending on the historical and current characteristics of the space, local uses and knowledge about the site, and community perceptions. These meanings depend in part on social-political climate and economic conditions locally and regionally. If the economy is growing, the availability of vacant lots as sites for capital investment and development is potentially an asset for a neighborhood, though it may also signal gentrification. But in declining urban economies, vacant spaces are problematic – the result of abandonment, disinvestment, and demolition. From merely 'vacant', these spaces can quickly become dangerous eyesores; empty lots turn into dumping grounds that are overgrown with weeds and rubbish, abandoned structures may serve as sites for criminal activities, and environmental quality is often very poor. In struggling economies such as Buffalo and many other deindustrialized urban areas, local governments are saddled with responsibility for these vacant spaces, so they often attempt to regulate vacant lots, offer them at tax sales, or compile inventories for potential investors. However, to the community members who live and work nearby, vacant lots may take on different meanings, based not on their potential capital worth, but on their social and community possibilities. Children may use the sites for exploration and play or as shortcuts through the neighborhood. Other community members may assume care of the redundant spaces in order to prevent unsavory activities and neighborhood decline. Local organizations may invest in cleaning up vacant sites for various purposes, including community gardens.

Throughout the history of the United States, vacant lots and vacant urban spaces have had different meanings and everyday connections with people, depending on context. The growing of food in community gardens is one common use for vacant spaces that has continually recurred in times and spaces of need. Government programs supporting vacant lot community gardening projects have emerged in economic recessions and during wars when resources were scarce, creating a space where citizens could grow their own vegetables and fruits. In times of economic depression, community gardening programs were intended to keep urban dwellers from becoming idle while producing food for their families in subsistence or allotment gardens (Hynes, 1996; Lawson, 2005). In wartime, growing food in community gardens known as victory gardens promoted conservation, nationalism, and patriotism, as well as redirecting gasoline and other resources for the military. However, as soon as the economy recovered or the war ended, government support for community gardens usually ended quickly too, and unless the local community instituted programs to retain the community gardens, most of the sites returned to their prior vacant state or were made available for capital investment (Lawson, 2005). These definitions all suggest that vacant spaces are *lacking*: lacking an occupant, a structure, a purpose, a caregiver, or a responsible owner. However, this vacancy may actually provide an opportunity for community members from marginalized groups to fill it with their will, their visions, their desiderata, through their everyday material and spatial practices and roles as community members and citizens.

More broadly, the struggle over vacant lots and the ensuing community gardens could be framed as a conflict over property rights and the right of the City, as the owner of record, to sell the lands as it saw fit. This struggle can also be viewed as a battle over 'the right to the city': over whose vision of the city should prevail; over exchange value versus use value; over the political and economic objectives of the urban elite versus the everyday needs of community members who live and work in the city.

For example, in New York City, residents created community gardens on vacant parcels that were the result of demolition, disinvestment, and abandonment during the fiscal crisis of the 1970s, but researchers found that once the economic conditions improved the City claimed that they had the right to sell the 'vacant' lots for development. The community gardeners also claimed 'ownership', based on several years and thousands of hours of sweat equity they had invested to create green spaces in a densely populated city with little access to open space (Staeheli et al., 2002). This small example of New York City's community gardens represents an excellent illustration of the ways that mixed methods can be used to develop a more complete understanding of the contested nature and social construction of spaces of the city. While places may be defined at a particular scale, for example at the scale of the parcel, they are *constituted* by processes that operate at multiple spatial scales. Staeheli et al. show how both spaces and spatial scales are contested, redefined, and restructured in ways that include community members, local government, and city officials within a multi-scale socio-economic context. Importantly for our discussion here, their work also demonstrates that discrete attribute categories of property value, zoning, and land use classifications such as 'vacant' do not capture the different variations of the discourse, the material realities, and the physical condition of vacant lots.

The varied meanings and uses of 'vacant' lots thus offer a rich terrain for exploring multiple realities, dismantling binaries, and performing reflexive data explorations. In order to put this study in its relevant context, we begin with a brief overview of the locale of Buffalo, New York.

Buffalo, NY

The story of Buffalo is a familiar one in the global North: rapid growth in the industrial period of the late nineteenth to the mid twentieth century, followed by fairly sudden factory closures and public disinvestment, net population loss, impoverishment of remaining residents (increasing proportions of whom are people of color), and the vast physical deterioration of public infrastructure, commercial and industrial properties, and private housing stock. In this sense, Buffalo, while not identical to other cities experiencing similar trends, has enough overlapping qualities with them to serve as a representative 'deindustrialized' or 'rust-belt' city. From a critical political-economy perspective, processes of geographically uneven development are inherently part of the downswing of capital investment, in that some places receive more investment while other places concurrently experience explicit under- or disinvestment in order to facilitate future rounds of capitalist development. Thus 'in times of economic crisis or restructuring when the built environment is no longer useful to production and/or consumption, substantial segments of the urban built environment may be abandoned' (Kaplan et al., 2004). The processes and outcomes of this abandonment are the focus here.

Today, Buffalo continues to face a declining tax base, loss of industry and jobs, increases in poverty and racial segregation, and a decrease in the central city population. These factors have resulted in a housing surplus and an oversupply of vacant and dilapidated housing stock in a city that was home to over 580,000 people in 1950, now reduced to a population of approximately 280,000 (*2005 Interim Census*; National Vacant Properties Campaign et al., 2006: 9). One report estimated that in 2003 there were 12,500 vacant residential, commercial, and industrial properties within the City of Buffalo, of which 4000 were city owned (Cornell Cooperative Extension, 2004). Burdened with the responsibility of maintaining the majority of these properties, the City had implemented an aggressive housing demolition policy, facilitating the demolition of over 2600 houses from 1998 through 2003 (Sommer, 2003). This context makes Buffalo an excellent case study for exploring the multiple meanings and definitions of 'vacancy', and illustrating some grounded visualization techniques in the process.

GROUNDED VISUALIZATION IN PRACTICE: RECURSIVE ANALYSIS OF VACANT LOTS AND/AS COMMUNITY SPACES

In the project examined here, several iterations of research were necessary to understand how community gardeners individually and collectively think about and use the material spaces of urban community gardens, particularly those located on 'vacant' lots, and how these processes emerged within a broader political and economic context. Official data (both quantitative and qualitative), including physical location, ownership type, assessed value, zoning, and other attributes of the community garden sites and the characteristics of the neighborhood, were essential to developing the research questions. But other primary research methods, including interviews, participant observation, photography, field inspections, and a survey questionnaire, were important for contextualizing the quantitative information and GIS to provide in-depth information about conditions, strategies, interactions, the processes of everyday life, and their relationships to larger structural conditions and processes. These different research practices were performed (by Knigge) simultaneously over the course of several years, but they were not done in *parallel*; rather, they constantly informed, contradicted, complemented, and enriched each other, which generated new questions and wove together new answers along the way. Thus, this project used an iterative, recursive process of integrating the analysis of these diverse forms of data, qualitative and quantitative, primarily through open-ended exploration and visualization of emerging themes, following new research directions that came up during the process, and allowing each round of knowledge building to inform the next. Although the research did not proceed in discrete pieces identifiable as 'visualization' and 'ethnography', we present the process and findings here from the angle of each of these two practices for clarity and to highlight their interaction.

Visualizing patterns of vacancy

In order to further understand the patterns of abandonment, vacancy, and demolition, 'official data' including census data, land use data, and parcel-level cadastral data were collected and mapped using GIS. This allowed me[2] to visualize the spatial distribution of the socio-economic and physical conditions of neighborhoods within the City of Buffalo.

Visual comparison of these maps revealed several patterns relevant to the research questions about the political and economic context that generates vacant land and the potential community uses for these spaces. First, a large, dense cluster of vacant lots on the East Side and a smaller cluster on the West Side of Buffalo raised questions about race and ethnicity because the main two racially marginalized communities (African American and Hispanic) are concentrated in these areas. Second, these two clusters also suggest questions about income levels: higher concentrations of vacant parcels were located in lower-income census tracts. Interestingly, majority white neighborhoods in South Buffalo (a long-standing Irish ethnic enclave) have fairly high rates of poverty, but very few vacant lots. Thus, the pattern of distribution of vacant land tends to be located in those areas with lower incomes and high rates of African Americans and Hispanics.

The four maps (in Figure 6.1) show the patterns of vacant land, median income, percentage of African Americans, and percentage of Hispanics in the city. The vacant land map depicts parcel-level data for the City of Buffalo, and the maps showing income and race/ethnicity use the 2000 US census data at the census tract level. The city's Office of Strategic Planning maps and datasets (OSP, 2004) were also useful for me in contextualizing and describing the patterns in more detail.

At one level, these maps and associated data further demonstrate that Buffalo is in the midst of an extended period of economic decline that is being unevenly experienced across the space of the city. This is evidenced by stark variations in median income, which are visible from the scale of the census tract up to the scales of the metropolitan area and the larger eight-county region of western New York State. The way that this decline is materially represented in the built environment varies within and between specific neighborhoods and districts, and at the metropolitan scale when comparing Buffalo to its suburbs. One of the most obvious impacts on the built environment is, of course, the massive loss of housing in some city districts.

To dig a little deeper here, I had discovered that most classifications of 'vacant' are the result of housing demolitions, which – as I found from newspaper accounts and community meetings – reflected a top priority of the mayor at the time. I also knew

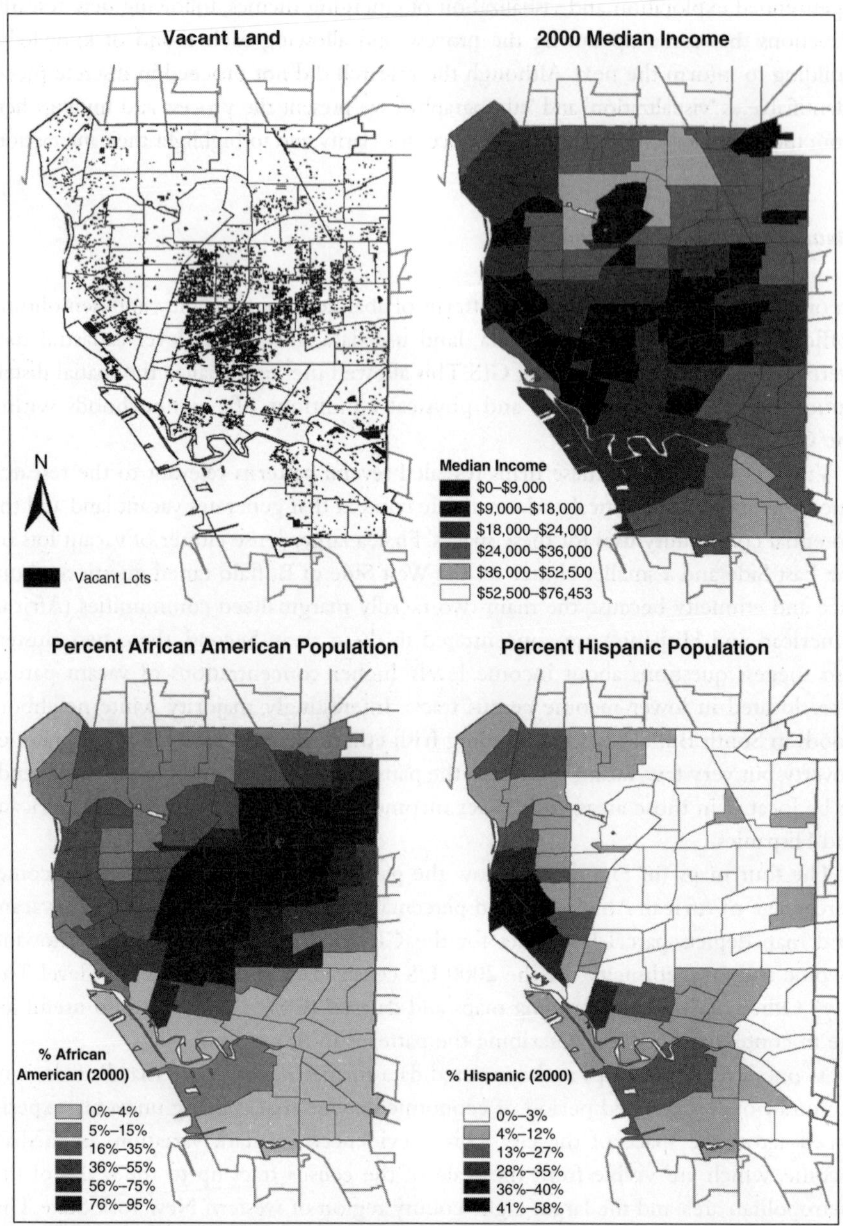

Figure 6.1 Visualizing patterns of vacancy in Buffalo, New York
(Source: 2000 US Census and Erie County parcel-level data)

from more general histories and demographic analysis (e.g. Kraus, 2000) that Buffalo has been the site of long-standing practices of racial segregation, much of which results from discrimination in the housing and job markets (Trudeau and Cope, 2003). The majority of the housing demolitions occur in the areas of Buffalo with the highest percentage of African Americans and Hispanics, whereas other areas of the city with low median income that are predominantly white experience only limited clusters of demolition. Clearly, at the scale of the city, the cumulative results of decades of race-related practices – both segregation and targeted demolitions – are apparent. Zooming into the census tract level, however, there seemed to be many different patterns. For example, there were a few tracts with very low residual populations as a result of extensive demolitions, but in other areas a spottier pattern of demolition had less of an effect on average tract characteristics. Thus, while there seemed to be some correlation between race/ethnicity and house demolitions, the visualizations also raised new questions for me about the hidden processes behind the patterns.

In reflecting on the visualizations and the questions sparked by these maps, broader critical questions about the production of knowledge also ensued. For example, all 'official data' are produced institutionally with a specific purpose, making certain claims to 'truth', and they represent one partial reality at a particular scale of aggregation. That is, data and cartographic representations created from them are imbued with the power of the institution that commissioned their collection, as well as the institution's agenda or purpose, and its scale of relevance. Census data are collected for particular purposes by powerful federal institutions and are discursively (re)produced by the rules, practices, and customs of those institutions, processes which themselves have been critically scrutinized by geographers. Hannah (2000) argues, for example, that the nineteenth-century origins of the national census in the US grew out of efforts to exert governmental power over not just population, but also territory, through both discourse and policy. Today, most official data are readily available for use by researchers and the public, yet their sheer volume and specificity can easily overwhelm users and preclude critical questions about classifications, binaries, and representations. In this way, grounded visualization can help critical scholars identify absences in the data, as well as question the discourse, power, and truth claims of the institutions that generate and selectively share data.

In the process of researching vacant lots and community spaces, the combination of these visualizations and contextual knowledge about Buffalo's racial histories and contemporary practices generated new questions. How did residents perceive these city-wide trends, and how did people's perceptions vary by neighborhood, race, and economic status? What were local people's views of increased vacancies in their neighborhoods or other sections of the city? Beyond perceptions of vacant spaces, did residents make use of the parcels in any way, and if so, how? These questions helped direct several iterations of ethnographic inquiry.

Ethnography and the processes of vacancy in community spaces

While my mapping and visualization efforts provided a representation of *some* material realities and spatial discourses within the City of Buffalo, they did not suffice for understanding the processes that generated and perpetuated vacant urban spaces. Additionally, I needed to understand the everyday experience of the community members who lived and worked within these spaces to consider how vacancy was mitigated, contested, and/or absorbed by local communities. I embarked on ethnographic field research to uncover some of these *other* material realities and discourses of space, and to consider the connections between official and community data, between definitions and meanings of vacancy, and between meaningful scales of analysis.

In the initial stages of the research I made numerous field studies of particular neighborhoods to gauge the impacts of vacancy at the neighborhood scale, not merely the parcel site itself. For example, one of the first field studies I completed involved making observations of conditions while walking the entire boundary of the Lower West Side neighborhood, referring frequently to the map of vacant parcels as I went. This revealed, among other things, that the use of '*vacant*' as a land use category in fact represented a vast continuum of conditions that ranged from veritable dumping grounds, overgrown with weeds and littered with trash, to beautiful, well-tended community gardens. As suggested above, the term 'vacant' implies a lack or absence, but field study observations revealed that many of the garden sites, far from suffering from lack, instead reflected the presence, action, and commitment of those who created the community gardens. The stark contrast between the unsightly, littered vacant lots and the beautiful community spaces created by the gardens led me to further question the definition of 'vacant' used by local government and land use planners and what these spaces meant to the community members who lived and worked there, inspiring further research questions. Who owned these parcels of land? Who cared for the gardens located on vacant lots? How long had they been there, and what was there before? Were the gardens open to the public or was access restricted? If the City planning office categorized them as vacant, was their use as a garden recorded or documented by the City?

To begin answering these questions, I engaged in many forms of observation of local neighborhood conditions by attending community meetings and garden tours, reading neighborhood newspapers and websites, tracking local issues in the main city newspaper, meeting with community leaders and gardeners, talking informally with residents, and documenting conditions through notes and photographs at dozens of vacant lots and community gardens at different times of year. I conducted formal and informal interviews with community members, representatives of community organizations, and local government officials and politicians. I also did some community mapping and distributed a survey that was completed by 91 community gardeners, representing 36 community gardens throughout the City of Buffalo. The interviews and survey questionnaires provide rich, contextual data about the community members, the gardens, their organization, what is done with the food that is grown, if any, and whether the organizations or individuals are involved with other forms of local community activism, among other things. Through the survey, I was able to find out specific information about the community gardeners such as where they were born; their age, race, ethnicity, and other demographic information; how long they had been involved with the community garden; their perceptions of their neighborhoods; and their knowledge of the history of the community garden sites.

I compiled all of these data into a community garden database that I maintained over five seasons to get a sense of change over time. The database also contained links to maps, visualizations, photographs, and spatially referenced quantitative data about the sites so that my analysis could draw simultaneously and recursively from multiple sources and types of data. This analysis led me to several themes that have relevance to the question of scale.

One of the first themes that emerged from the research involved the ways that community members are frequently called upon to care for the spaces of their neighborhoods. For example, at the first meeting of the Buffalo Coalition of Community Gardeners and Vacant Lot Task Force that I attended, a variety of people including members of block clubs and other community organizations, academics from local universities, city employees, a city councilman, representatives from youth groups, and local residents were in attendance. Members of the community in attendance expressed several concerns with vacant lots within the city such as the mowing and maintenance of these lots, possible soil contamination and soil remediation techniques, the lots' assessed value, the potential for urban agriculture, and restrictive ordinances against livestock in the city. Many of the attendees of the meeting were both frustrated with the lack of support and interested in building coalitions of support with city government and its representatives. The representatives from city government were receptive to collaboration; frequently, in times of economic hardship, cities readily relinquish control, use, and care of idle, redundant spaces under its authority to community members to maintain until economic conditions improve. In 'allowing' community members to take over the material use of these vacant spaces, the city (or absentee owner) is relieved of the responsibility and obligation to care for the properties.

While community members use this vacancy in ways that fill their needs – and, indeed, *fill* the empty spaces of the lots – their tenure is contingent and considered an interim use by the city. Cities in these circumstances are often caught between promoting, packaging, and 'selling' the city to potential capital investors and the need to maintain the city for the residents that live and work there. This suggests ways that the scale of concern for local residents differs from the scale of concern for the city: residents in this study cared most about their block, neighborhood, or district, while the typical political-economic pressures on city managers result in a focus on regional, state, or even global competitive place marketing. Thus, city government is often in the position of a scale mediator, trying to balance the localized issues of individual parcels of land (and the desires and visions of community residents for that land) and the much broader-scale concerns of attracting capital investment from global and national directions. The following case study exemplifies this process and these tensions, but also serves to demonstrate the iterations of research – both ethnography and visualization – that were involved in building on the emerging theme of the City as a scale mediator.

The City as a scale mediator

Connecticut Street, once a vibrant, commercial street on Buffalo's West Side, had fallen on difficult times. From mapping census data I had found the area had high rates of poverty, and a transitional demographic as older white ethnic groups (such as Italians and Eastern Europeans) aged out or left for inner-ring suburbs, to be replaced in the 1990s and early 2000s by African Americans, Puerto Ricans, and several other new immigrant and racial/ethnic groups. From my field studies I had noted many empty store fronts, vacant lots, and dilapidated housing on and around Connecticut Street. And from a review of newspapers and attendance of community meetings, it was clear to me that the area had problems with gangs, vandalism, and graffiti.

Beginning in 2003 I was able to document several new developments among local residents. The Connecticut Street Neighborhood Association instituted a graffiti abatement program, held neighborhood cleanups, implemented greening strategies that included planting trees and a community gardening program, and instituted a push-cart vendor initiative to create local economic development. In June 2003, the first New World Street Market Connecticut Street Festival, attended by over 5000 people, was part of the continued efforts to revitalize Connecticut Street. Community gardens, sponsored by various community organizations and local businesses, were planted on four of the numerous vacant lots located on the street with labor from community volunteers and the YO! Buffalo youth group. I recorded these developments through photographing sites, talking with activists and resident gardeners, attending events and community meetings, and tracking newspaper coverage.

I found that through their activities, community members were actively engaged in the production of social and financial capital at a local scale, both materially through their interventions on the landscape of Connecticut Street and discursively through their engagement with media, government decision makers, funding agents, community organizations, and local residents themselves. While City of Buffalo policies *allowed* community members the material use of the vacant lands through contingent leases of vacant properties for community gardens, ultimately the goal of the City was to prepare the vacant spaces for sale, toward stimulating capital (re)investment and the parcels' return to the tax rolls. Thus, the public is permitted to aid the City in this process through interim uses if it is of benefit to both the immediate community and the City. However, tensions and conflicts inevitably arise.

When the nearby site of the Hudson-West Community Garden in the Lower West Side was to be sold by the City for infill public housing under Hope VI (a program of the federal US Department of Housing and Urban Development) in 2005, community members became mobilized. The garden had recently been taken over by the Nickel City Housing Co-op because the previous community garden organizer had become ill. When news of the possible sale of the site for federally funded housing was announced, the members of the Housing Co-op went door to door in the neighborhood with a petition supporting the garden, organized community meetings, contacted the local city councilor and other politicians, and were able to garner enough support for the community garden that the plan to develop the site for Hope VI housing was abandoned. Admittedly, the City and the developer had little to lose: public housing does not result in much (if any) tax revenue, and the developer had plenty of other vacant sites to choose from in the city. But the important point is that the community gardeners became materially and discursively engaged in the production of their scale of concern – the neighborhood – through their social and political actions, as well as through their everyday actions in tending the community garden.

An interesting counterpoint to this happened the following year. In 2006, one of the previously mentioned community garden sites on Connecticut Street was purchased from the City and a Greater Buffalo Savings Bank was built on the site. While the loss of a community garden was a mobilizing event in some places and at some times, in the case of Connecticut Street, community members did not challenge or contest the sale of this garden site for the construction of the bank; rather they welcomed the opportunity for economic development. This example is important in demonstrating the value of familiarity with the context of neighborhood change and community mobilization. In particular, it suggests that the community itself did not necessarily envision only one appropriate solution to vacant lots (community control over gardens or other community-focused spaces) but, rather, welcomed for-profit capital investments that stimulated the neighborhood's economy in other ways. At one level, the fact that the community garden had occupied the space for some time made it more attractive to the bank, which could be seen as exploiting the efforts of the gardeners, but at another level the gardeners' investments of time and money paid off for the neighborhood as it gained an important local institution and a needed service for residents. These kinds of episodes again demonstrate the nuanced understandings that can emerge best from in-depth, integrated research, and mixed methods approaches such as grounded visualization are of particular assistance.

CONCLUSION: GROUNDED VISUALIZATION AS INTEGRATED, REFLEXIVE SCALE ANALYSIS

During the course of this research, it had become very clear that decisions regarding housing demolitions and the sale of vacant lots were inherently political, as well as economic and bureaucratic. In interviews and casual conversations, many residents spoke of requesting that the City demolish certain derelict properties and expressed frustration and anger over the slow pace at which housing demolitions occurred. Others had sought injunctions halting the demolition of historic houses. Some grassroots community organizations wanted to increase the rate of home ownership and housing rehabilitation efforts and decrease the rate of housing abandonment.

The ethnographic methods I used elicited a set of processes, practices, and preferences that were nuanced, based on highly localized context, and often reflected conflicts both between residents and the City and within the community itself. These insights were richly detailed and helped in understanding the meanings of 'vacancy' on the ground, which then reflected back into analyses of the patterns of vacancy in visualizations of official data. Similarly, the ethnography revealed the social, economic, and political processes of the construction of scale as the cumulative result of ongoing, contested, and differentiated decisions made by actors and agents with divergent views and goals.

In many studies qualitative research is seen as providing context for the patterns that are generated with quantitative data visualizations, but this case illustrates that the reverse is also true: that visualizations can provide context for ethnographic research.

For instance, in an interview with an East Side community activist who had organized numerous gardens, directed a community center, and challenged City officials repeatedly, the context of the social, economic, and physical conditions in her neighborhood *mattered* greatly for my interpretation of her words. Significantly, I was familiar with this context – in this case, the Ellicott District – through the research and visualizations I had done with official data: many blocks located in the East Side of Buffalo have experienced so many demolitions that lone houses stand in the middle of vast parcels of vacant land where single family homes once stood; 60% of the Ellicott District population is black or African American, 25% is white, and 17% identify as Hispanic; 39% of the residents of the Ellicott District live below the poverty level with an average household income of $29,183, the second lowest of all of districts in Buffalo; and the vacant housing rate in the Ellicott District is 24%, one of the highest in the city (Buffalo Planning Analysis/GIS, 2003). Being familiar with these local conditions influenced my interaction with this activist on the day of the interview, but also shaped the ways that I interpreted the woman's words later in the analysis, which in turn led me to new interpretations of the maps and quantitative data for the Ellicott District.

These examples from the West Side and the East Side of Buffalo help to instantiate the three dimensions of scale in grounded visualization that we laid out above. First, the scale at which *data* were collected/produced for official purposes and for scholarly research revealed multiple realities (in this case, meanings and definitions of 'vacancy'), which were themselves often in conflict. Second, the *processes* of creating, managing, and mitigating the effects of vacant lots, as conducted both by the City and by local residents and organizations, are inherently related to the construction of the 'scale of concern' for different agents, which in turn fosters a more critical look at the (superficially) binary category of 'vacancy'. Finally, multiple representations of official data at different levels of resolution and with different layers of data shown or hidden allowed Knigge to engage in exploratory analysis not just with the visualizations themselves, but also in concert with the ethnographic findings. Through the lens of scale, then, these analyses demonstrate more fully the explorative, iterative, and reflexive capacities of grounded visualization.

NOTES

1 We use the term 'official data' to refer to census data, land use data, cadastral parcel-level data, and other forms of secondary data collected by the Census Bureau, city and county government, and other state sponsored entities. These spatially referenced data can be mapped and explored using geographic information systems (GIS) and other computer programs and are readily available to the public for data analysis and manipulation.

2 The shift to the first-person voice here represents the fact that this section is based on Knigge's research process, experiences, and strategies within the case study project.

REFERENCES

Brewer, C. and Suchan, T. (2001) *Mapping Census 2000: The Geography of US Diversity*. US Census Bureau, Census Special Reports, Series CENSF/01–1. Washington, DC: US Government Printing Office.

Brown, M. and Knopp, L. (2006) 'Places or polygons: governmentality and sexuality in *The Gay and Lesbian Atlas*', *Population, Space, and Place*, 12 (2): 223–42.

Buffalo Planning Analysis/GIS (2003) *Census 2000 SF3 Summary Statistics: Erie County, City of Buffalo and Buffalo Council Districts*. Buffalo, NY: City of Buffalo.

Charmaz, K. (2000) 'Grounded theory: objectivist and constructivist methods', in N.K. Denzin and Y.S. Lincoln (eds), *Handbook of Qualitative Research*. Thousand Oaks, CA: Sage. pp. 509–36.

Cornell Cooperative Extension (2004) *Vacant Land, Buildings and Facilities Asset Management Project: A Project Report*. Buffalo, NY: City of Buffalo, Buffalo Urban Renewal Agency, Cornell Cooperative Extension Association of Erie County, Cornell University Cooperative Extension–Community and Economic Vitality, Cornell University Community and Rural Development Institute.

Elwood, S. (2009) 'Mixed methods: thinking, doing, and asking in multiple ways', in D. DeLyser, M. Crang, L. McDowell, S. Aitken and S. Herbert (eds), *The Handbook of Qualitative Research in Human Geography*. London: Sage.

Fiedler, R., Schuurman, N. and Hyndman, J. (2005) 'Improving census-based socioeconomic GIS for public policy: recent immigrants, spatially concentrated poverty and housing need in Vancouver', *ACME: An International E-Journal for Critical Geographers*, 4 (1): 145–71.

Haining, R. and Wise, S. (1997) 'Exploratory spatial data analysis', *NCGIA Core Curriculum in GIScience*. Santa Barbara, CA: NCGIA. http://www.ncgia.ucsb.edu/giscc/, accessed 30 June 2008.

Hannah, M. (2000) *Governmentality and the Mastery of Territory in Nineteenth Century America*. Cambridge, UK: Cambridge University Press.

Harley, J. (1989) 'Deconstructing the map', *Cartographica*, 26 (2): 1–20.

Hynes, H. (1996) *A Patch of Eden: America's Inner-City Gardeners*. White River Junction, VT: Chelsea Green.

Kaplan, D., Wheeler, J. and Holloway, S. (2004) *Urban Geography*. New York: Wiley.

Knigge, L. and Cope, M. (2006) 'Grounded visualization: integrating the analysis of qualitative and quantitative data through grounded theory and visualization', *Environment and Planning A*, 38 (11): 2021–37.

Kraus, N. (2000) *Race, Neighborhoods, and Community Power: Buffalo Politics, 1934–1997*. Albany, NY: SUNY Press.

Lawson, L. (2005) *City Bountiful: A Century of Community Gardening in America*. Berkeley, CA: University of California Press.

Marston, S. (2004) 'A long way from home: domesticating the production of scale', in E. Sheppard and R. McMaster (eds), *Scale and Geographic Inquiry: Nature, Society, and Method*. Malden, MA: Blackwell. pp. 170–91.

McDowell, L. (1999) *Gender, Identity and Place: Understanding Feminist Geographies*. Minneapolis: University of Minnesota Press.

McDowell, L. and Sharp, J. (1997) *Space, Gender, Knowledge: Feminist Readings*. London: Arnold.

McLafferty, S. (2002) 'Mapping women's worlds: knowledge, power, and the bounds of GIS', *Gender, Place, and Culture*, 9 (6): 263–9.

McMaster, R., Leitner, H. and Sheppard, E. (1997) 'GIS-based environmental equity and risk assessment: methodological problems and prospects', *Cartography and Geographic Information Science*, 24 (3): 172–89.

Moss, P. (2003) *Feminist Geography in Practice: Research and Methods*. Malden, MA: Blackwell.

National Vacant Properties Campaign, Local Initiatives Support Corporation–Buffalo, Schilling, J., Schamess, L. and Loga, J. (2006) 'Blueprint Buffalo Action Plan: regional strategies and local tools for reclaiming vacant properties in the city and suburbs of Buffalo', in National Vacant Properties Campaign (ed.), *Creating Opportunities from Abandonment*. Washington, DC.

New York State Office of Real Property Services (2006) *New York State Department of Real Property Services Assessor's Manual*. Albany, NY.

OSP (2004). *Census 2000 SF3 Summary Statistics: Erie County, City of Buffalo, and Planning Communities and Neighborhoods*. Office of Strategic Planning, Buffalo, NY. http://www.ci.buffalo.ny.us/files/1_2_1/SPlanning/Census2000SF3.pdf, accessed 13 June 2008.

Sheppard, E. (1990) 'Ecological analysis of the "urban underclass": commentary on Hughes, Kasarda, O'Regan and Wiseman', *Urban Geography*, 11 (3): 285–97.

Slocum, T.A., McMaster, R.B., Kessler, F.C. and Howard, H.H. (2009) *Thematic Cartography and Geographic Visualization*, 3rd edn. Upper Saddle River, NJ: Pearson Prentice Hall.

Sommer, M. (2003) 'City's house demolitions add up to a mixed blessing', *The Buffalo News*, 17 March: Aa.

Staeheli, L., Mitchell, D. and Gibson, K. (2002) 'Conflicting rights to the city in New York's community gardens', *GeoJournal*, 58 (2–3): 197–205.

Trudeau, D. and Cope, M. (2003) 'Labor and housing markets as public spaces: "personal responsibility" and the contradictions of welfare-reform policies', *Environment and Planning A*, 35 (5): 779–98.

Women and Geography Study Group (1997) *Feminist Geographies: Explorations in Diversity and Difference*. Essex: Addison Wesley Longman.

7

COMPUTER-AIDED QUALITATIVE GIS: A SOFTWARE-LEVEL INTEGRATION OF QUALITATIVE RESEARCH AND GIS

Jin-Kyu Jung

ABSTRACT

A growing number of geographers are conducting mixed methods research involving the integration of quantitative and qualitative data in GIS. Contributing to these efforts, this chapter describes software-level modifications that adapt GIS to enable inclusion of qualitative data as well as interpretive codes associated with these data. These innovations enable GIS to serve as a platform for dynamically integrating quantitative and qualitative data throughout the analysis process. Further, this chapter shows how GIS may be meshed with computer-aided qualitative analysis software (CAQDAS) to support inductive interpretive analysis. The value of GIS is in its ability to represent both qualitative and quantitative data along with their spatial information, and the value of CAQDAS lies in its ability to provide better means of storing, managing, and analyzing qualitative data. The system described here enables researchers to take advantage of all of these capabilities as they are working with multiple forms of data. Further, the linkage between GIS and CAQDAS that I have developed enables researchers to carry out many different forms of analysis, such as exploratory data visualization, conventional forms of spatial analysis, grounded theory, and other approaches.

INTRODUCTION

GIS is commonly used to display and analyze urban demographic information such as population, racial distributions, and median income, and such GIS-based data are widely used by policy makers for making plans and decisions that affect urban neighborhoods. But none of these data enable us to explore local people's individual lived experiences, or their attachments and social ties to their neighborhoods. Such neighborhood-level local knowledge can offer researchers and policy makers alike 'well-grounded, rich descriptions, and explanations of processes in identifiable local contexts' (Miles and

Huberman, 1994: 1). Researchers in public participation GIS (PPGIS) have developed ways of incorporating local knowledge into GIS-based maps and facilitating its inclusion in planning processes that use GIS (Aitken and Michel, 1995; Al-Kodmany, 2002; Barndt, 1998; Corbett and Keller, 2005; Elwood, 2006; Elwood and Leitner, 1998; Ghose, 2001; Ghose and Huxhold, 2001; Harris and Weiner, 1998; Weiner and Harris, 2003). But these researchers and others illustrate the challenge of including local knowledge in GIS-based data structures and GIS-based analysis, because it is often represented in forms such as texts, image, audio, and video, rather than numbers or maps. The project described in this chapter addresses this challenge, developing ways of incorporating well-grounded contextualized qualitative data, which reflect the everyday lives of people and society, directly into geographic information systems. More broadly, this project contributes to qualitative GIS methodologies by developing ways to both include qualitative forms of data and carry out qualitative analysis techniques. These kinds of qualitative GIS practices produce new ways of using GIS in research, leading us to a form of mixed methods practice that weaves together GIS and qualitative research.

A growing number of geographers are using GIS as part of mixed methods research that incorporates both quantitative and qualitative forms of data and analysis (Gilbert and Masucci, 2005; 2006; Kwan and Knigge, 2006; McLafferty, 2005a; 2005b; Schuurman and Pratt, 2002; Sheppard, 2001). Building on these approaches, I have developed a new approach that integrates GIS with qualitative data and computer-aided qualitative analysis, a technique I refer to as computer-aided qualitative GIS (CAQ-GIS). This chapter presents the key concepts and specific software innovations through which I implement this new approach for blending qualitative research with GIS. In the second section, I conceptualize the 'qualitative' in 'qualitative data', drawing on existing literature on critical visual and qualitative methodologies, and review existing approaches for handling qualitative data in GIS. I argue that there are two components to a robust qualitative GIS which existing approaches have not yet fully integrated: incorporating qualitative data into GIS data structures, and supporting qualitative analysis techniques that can be brought to bear upon these data. In the third and fourth sections, I present the software innovations I have developed to achieve these goals. The system I have developed, CAQ-GIS, enables storage of qualitative data such as images directly into GIS data structures, and facilitates inductive interpretive analysis of these data using GIS software and a computer-aided qualitative data analysis software (CAQDAS). I demonstrate the capabilities of this integrated system using examples drawn from an ethnographic research project on children's urban geographies carried out in Buffalo, New York.

The value of GIS is in its ability to represent both qualitative and quantitative data along with their spatial information, and the value of CAQDAS lies in its ability to provide a means of storing, managing, and analyzing qualitative data. With the techniques I have developed, GIS and CAQDAS are integrated with each other, enabling the researcher to take advantage of all of these capabilities in a systematic and coordinated way. The system I have developed is intended to support innovative mixed methods research, developing software-level structures that will support researchers' efforts to integrate spatial analysis, qualitative visualization, and other qualitative analysis techniques.

EXISTING CONCEPTUALIZATIONS AND STRATEGIES FOR SUPPORTING QUALITATIVE VISUALIZATION IN GIS

An important first step toward developing qualitative approaches to GIS is conceptualizing precisely what is meant by 'qualitative' and how and where this 'qualitative-ness' is present: in spatial data, in visual representations of these data, in data analysis techniques, or in other components and processes of GIS. In this project, I focus upon incorporating qualitative data in GIS and facilitating qualitative analysis with GIS software. In doing so, I conceive of qualitative data as those forms of information that provide well-grounded, richly contextual descriptions of everyday spaces and activities, collected through research methods such as interviews, oral histories, sketch mapping, photography, or drawing. That is, I understand qualitative data to be more than simply those data that are non-numerical. Rather, qualitative data are representations of the experiential knowledge of individuals or groups, and the interpretations and meanings they ascribe to these representations.

Researchers in critical cartography and critical geography more broadly have shown how visual representations such as maps are socially embedded and constructed, laden with interpretive contextual meanings (Crampton and Krygier, 2006; Dorling, 1998; Fiedler et al., 2006; Harris and Harrower, 2006; Harris and Hazen, 2006; Johnson et al., 2006; Kanarinka, 2006; Krygier, 2006; Wood, 2006). Writing on critical visual methodologies, Rose (2001) notes that images are made and used in all sorts of ways by different people for different reasons. Or, as Krygier notes, 'Visualization is not arbitrary but more concrete because we have a point to make, a story to tell, and knowledge to communicate' (1999: 48). Following from these perspectives, *any* image may be qualitative, if it encodes and visualizes qualitative information. Further, a GIS-based map can be seen as an image that produces and represents qualitative information. This perspective opens the door to considering many ways that qualitative information may be included in a GIS by including images.

But it is critical to note that not all images function as qualitative data. A satellite image or air photo, for example, is often considered a reflection or exact mirror of the 'real world', perhaps detailing the physical structure of a neighborhood. In an urban GIS application, such an image might be used to show the presence or location of streets, buildings, or parks, or to measure and check the accuracy of other spatial data files. These types of images, which I refer to here as 'outlook images', do not function as qualitative data, because they do not allow us to discern anything about social relations, different experiences of places, or meanings and identities that different actors may attach to these places.

In contrast, 'qualitative images' are those that convey multiple meanings of a place, such as neighborhood or community.[1] Qualitative images may be maps, photographs, sketches, or other visual forms of information that carry the interpretations, meanings, and experiential knowledge of the individuals or groups who created them. For example, Figure 7.2 includes some photographs of the West Side neighborhood of Buffalo, New York. I created these images when I visited this area as a researcher. The photographs are not simply randomly selected snapshots. Rather, each is an image I recorded as an exemplar of key aspects of the neighborhood environment. So the photograph is already a representation that tells a story. Different stories and meaning might be conveyed if different researchers or local residents took photographs for inclusion in the GIS. To differentiate this type of image from 'outlook images', I refer to them as 'neighborhood images'.

As scholars writing on critical visual methods have shown, interpreting the meanings associated with such qualitative images is challenging, particularly in efforts to interpret meanings signified in an image created by another person (Kwan, 2002a; Rose, 2001). Using images in a GIS as a means of incorporating qualitative information will of course carry these same difficulties. But more immediately, there is the practical challenge of how to incorporate such qualitative knowledge in digital form in a GIS. Several approaches have been developed by other researchers. Some focus upon including local knowledge, such as unique locally specific place names, into a thematic layer in the GIS (Flavelle, 1995). Others rely on a multiple methods research design, using GIS-based analysis and representation in combination with other forms of data analysis and representation, instead of making qualitative data and analysis part of the GIS itself (Dennis, 2006; McLafferty, 2002; Pavlovskaya, 2002; 2006).

Still other scholars use the visualization capabilities of GIS to integrate or link multiple forms of evidence with GIS maps and the objects shown in them. Pavlovskaya (2002), for example, in a study of urban economies in post-Soviet Moscow, used information such as census data to characterize the formal monetary economy, and information household interviews to identify key sites and characteristics of the informal non-monetary economy. Both kinds of evidence were represented in GIS-based maps. But the information drawn from the household interviews was not incorporated directly. Rather, this narrative information was used to identify key sites and their characteristics, which were then encoded in the GIS database and represented on the map. This approach involves a transformation of primary data, a sort of 'quantification' of qualitative information so that it might be visualized in tandem with other forms of evidence such as the census data.

Pavlovskaya (2002) shows how incorporating multiple forms of evidence greatly strengthens the explanatory power of her GIS-based analysis. However, the process of quantifying qualitative information has limitations. Specifically, in transformation of original data, much of the rich contextual interpretive knowledge communicated in the interview narratives is surely lost. Quantifying qualitative data for visualization is only a partial solution, because only a portion of the 'qualitative-ness' of the knowledge gathered from households is included.

One way of avoiding the problems posed by such transformations is to include the original data directly, and connect it to the GIS. The most common approach for doing so is using hyperlinks or hypermedia tools to link multimedia data such as text, images, audio, and video to spatial objects represented in a GIS. For example, Al-Kodmany (2000) uses multimedia approaches to include panoramic photographs and video clips of a neighborhood in a GIS as part of participatory environmental design projects. Hyperlinking techniques are an effective way to incorporate multiple forms of knowledge and have been successfully used in many participatory GIS applications to represent diverse experiences, individuals, and social groups (Al-Kodmany, 1999a; Cieri, 2003; Krygier, 1999; 2002; Talen, 1999; 2000; Weiner and Harris, 2003). By including such qualitative data directly, we might be able to preserve the contextual interpretive meanings they contain. But an important limitation of hyperlinking strategies is that the qualitative data are stored outside the GIS database. The hyperlinks simply associate these multimedia data with particular geographic entities that are represented in the GIS (such as a river, a house, or a forest). In this approach, the qualitative data are not georeferenced: they do not contain any explicitly spatial information such as latitude/longitude coordinates. Thus, the qualitative data can be used for GIS-based visualization, but it is difficult to incorporate these data in any other analysis within the GIS, or even to query and retrieve them based upon their location, because these data are not part of the GIS itself.

In sum, these existing strategies have provided important tools for linking qualitative data with GIS. They primarily incorporate qualitative data by visually representing them on the map or alongside the map. But maps are only one component of a GIS. Much of the analytical power of GIS is vested in query, selection, reclassification, and other operations performed upon the data stored in these data structures of the GIS itself. If qualitative data are to be the target of GIS-based analytical operations, they must be encoded directly into the database. In hyperlinking and other visualization-based efforts to include qualitative data in GIS, the data remain external to the GIS. But if qualitative data can be stored directly in GIS data structures, we can carry out analysis of these data without relying on such practices as quantifying qualitative data. Encoding qualitative data in GIS data structures also opens the door to using GIS in concert with qualitative analysis techniques, rather than using GIS only for visualization of multiple forms of data at the final representational stage of research. In the following section, I describe the software-level innovations I have developed to support the inclusion of qualitative data directly into GIS data structures.

ADAPTING GIS DATA STRUCTURES TO STORE QUALITATIVE DATA: THE IMAGINED GRID AND THE HYBRID RELATIONAL DATABASE

My approach for incorporating qualitative data, such as the 'neighborhood images' described in the previous section, into a GIS builds upon the existing data models of a GIS. There is untapped potential in these models for incorporating qualitative data directly into GIS data structures. As Schuurman (2000; 2004) explains, GIS represents geographic objects and their characteristics through an object-based ontology or a field-based ontology. These ontologies are implemented in GIS software through either a vector model or a raster model. In a vector model, geographic objects (such as roads, cities, or conservations areas) are represented using three geometric primitives: points, lines, and polygons. Then, attribute data describing the characteristics of these objects (road type, city population, species present in the conservation area) are associated with the object. Typically, this association is made through a relational database in which some tables define the geographic objects (their locations, shape, size, and so on), and others store the attribute data. The tables are related through some unique identifiers assigned to each entity. A raster data model defines an area through a series of grid cells, and then defines geographic objects within this area by assigning values to each cell. Thus, a geographic object such as a forest is represented by encoding multiple adjacent cells with some value that is assigned to represent 'forest'.

As Schuurman (2000; 2004), Chrisman (1997), and others have illustrated, both raster and vector models establish systematic approaches to representing geographic objects and their characteristics in a digital environment, and facilitate analysis techniques based upon Boolean logic and map algebra. The limits of relying solely upon these approaches have been well explained by Sheppard (1995; 2005). But I find that strategies used in implementing vector and raster models can be adapted to support the storage and analysis of qualitative data. Specifically, I rely upon the raster model's grid structure and the vector model's relational tables to develop two new innovations, which I term the 'imagined grid' and the 'hybrid relational database'. Below I describe these structures and how they might be used in qualitative GIS approaches.

To incorporate qualitative data directly into GIS data structures, my strategy is to create a special layer for storing qualitative data, using techniques from existing data models in GIS. Creating a special layer for qualitative data solves several of the limitations posed by hyperlinking approaches, such as external storage and the absence of spatial identifiers for the qualitative data. I use strategies from the raster model to create this special layer for storing qualitative data. As shown in Figure 7.1, this layer, called the 'imagined grid', comprises regular grid cells overlaying the other data layers in use in the GIS.

Once the imagined grid is defined, qualitative data such as neighborhood images can be stored in its grid cells. However, before the data can be inserted in the imagined grid, they must be georeferenced. I accomplish this by giving each neighborhood image the coordinates from the grid cell to which it is assigned. In the imagined grid, as in any raster, the vertices of each grid cell have coordinates that define their geographic location. These same coordinates may be assigned to the four corners of the neighborhood image that will go into that grid cell.[2] Once each neighborhood image is georeferenced and rectified through this process, location information is associated with it, allowing it to be incorporated directly into the imagined grid. Figures 7.2 and 7.3 illustrate what this approach looks like, when images stored in the imagined grid are represented on the map, together with other data layers. These figures show neighborhood images that were stored in the imagined grid, displaying them together with other forms of data: a layer showing streets; a satellite image of the area in Figure 7.2; and in Figure 7.3 a thematic map created with demographic data from the US Census. These data are from the case study I conducted in the West Side neighborhood of Buffalo, New York.

Figures 7.2 and 7.3 begin to illustrate some of the advantages of using the imagined grid to store qualitative data. Because qualitative data are stored in the GIS attribute table, they can be more easily retrieved and used at any stage of the research, especially in analysis. With this approach, qualitative images of neighborhood can be incorporated with other data, both quantitative and qualitative, such as census demographic data, seasonal images, and multi-scale visual data in GIS (Figure 7.1C). Using the imagined grid to include qualitative data, a user could simultaneously visualize multiple forms of evidence, exploring relationships and patterns within and across different data. Such an approach might be particularly useful for a researcher carrying

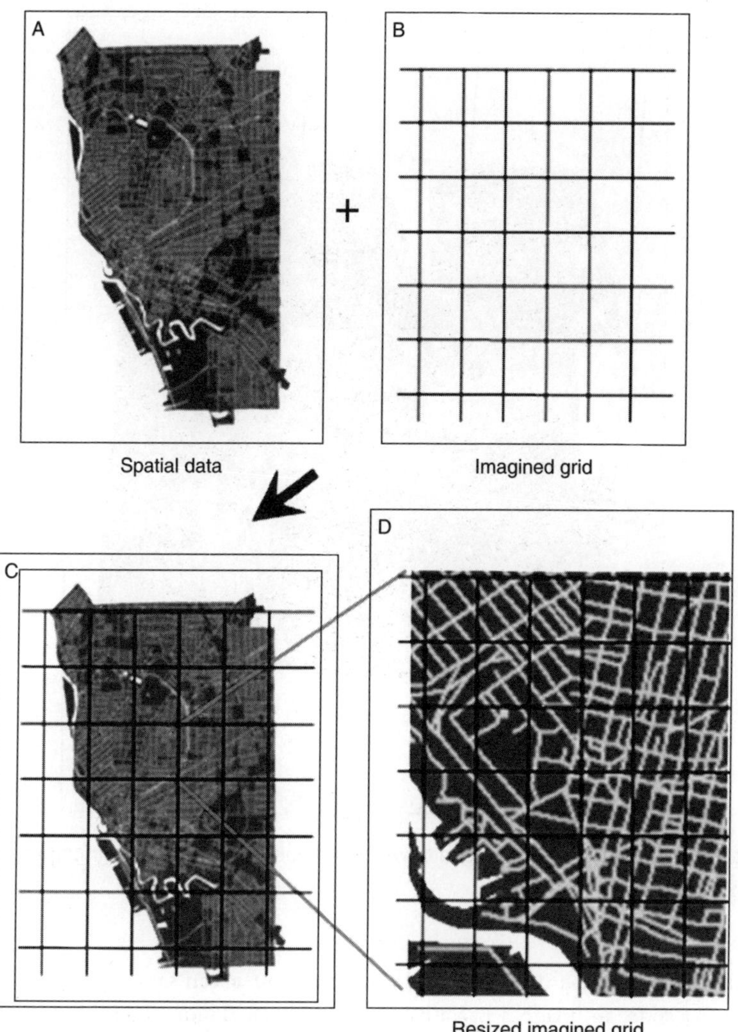

Figure 7.1 The imagined grid

out Knigge and Cope's (2006) grounded visualization technique, an iterative inductive analysis of multiple forms of evidence in a GIS. Another advantage of the grid, practically and technically, is its scalability. As shown in the 'resized imagined grid' in Figure 7.1D, if we change the scale of the map, the neighborhood images are automatically adjusted like other geographical spatial data in the GIS.

Using concepts from the raster model, but adapting them to store qualitative data, the imagined grid is an improved way to collect, store, manage, and visualize qualitative data in GIS. But the imagined grid is still not sufficient alone. It enables

Figure 7.2 Using the imagined grid to visualize qualitative images: neighborhood images displayed with satellite imagery

inclusion of qualitative data into a GIS database, but it can save only one qualitative artifact – a code, an image, a narrative – per grid cell (Figure 7.4A). It would not be possible, for example, to store both a qualitative image and an interpretive code assigned to that image by a researcher in the process of analysis, because there is no way to save more than one piece of information in each cell. This limitation is inherent in raster models. We could create multiple imagined grids to solve this problem, but this is a very inefficient solution in terms of data storage and computation.

I solve this problem with the relational database structure that is commonly used in a GIS to link tables storing attribute data to tables that define the geographic objects in a vector model. The relational database structure enables us to build one-to-many relationships, so that a single record (row) in one table might be associated with multiple records (rows) in another table, as shown in Figure 7.4.

Through relational database structures, we can associate more than one qualitative data attribute with a single grid cell in the imagined grid. First one creates multiple records in the attribute table (called the 'qualitative data table' in Figure 7.4B). Then these records can be used to contain multimedia qualitative data such as images, text,

Figure 7.3 Using the imagined grid to visualize qualitative images: neighborhood images with thematic map of demographic data

audio, or video. Additional fields (columns) in this table could be populated with interpretive codes created by the researcher in the process of analysis, as shown in the columns labeled 'Code1' and 'Code2' in Figure 7.4B.

I call this design a 'hybrid relational database'. A hybrid relational database consists of an imagined grid (identified as the 'spatial data table' in Figure 7.4A), plus the relational table that contains original/raw qualitative data or interpretive codes (identified as the 'qualitative data table' in Figure 7.4B). The qualitative data table is relationally joined to the spatial data table using a unique identifier. As shown in Figure 7.4, the qualitative data table can associate multiple records (rows) with a single record/object in the spatial data table. This design is 'hybrid' in several ways: it can incorporate qualitative data and quantitative data, it can contain multiple media, and it can contain 'raw' data and interpretive analytical information from the researcher.

Together, the imagined grid and the hybrid relational database can support visualization and inductive analysis of multiple forms of data. To demonstrate this capability, I created these structures in ESRI's ArcGIS 9.2 while participating in a research project

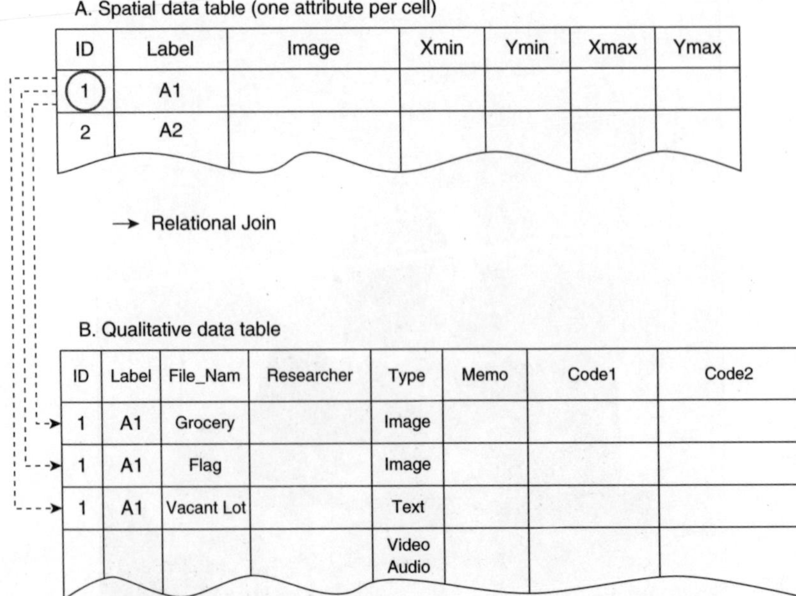

A. Spatial data table (one attribute per cell)
ID
Label
Image
Xmin
Ymin
Xmax
Ymax
1
A1
2
A2
Relational Join
B. Qualitative data table
ID
Label
File_Nam
Researcher
Type
Memo
Code1
Code2
1
A1
Grocery
Image
1
A1
Flag
Image
1
A1
Vacant Lot
Text
Video
Audio

Figure 7.4 Hybrid relational database

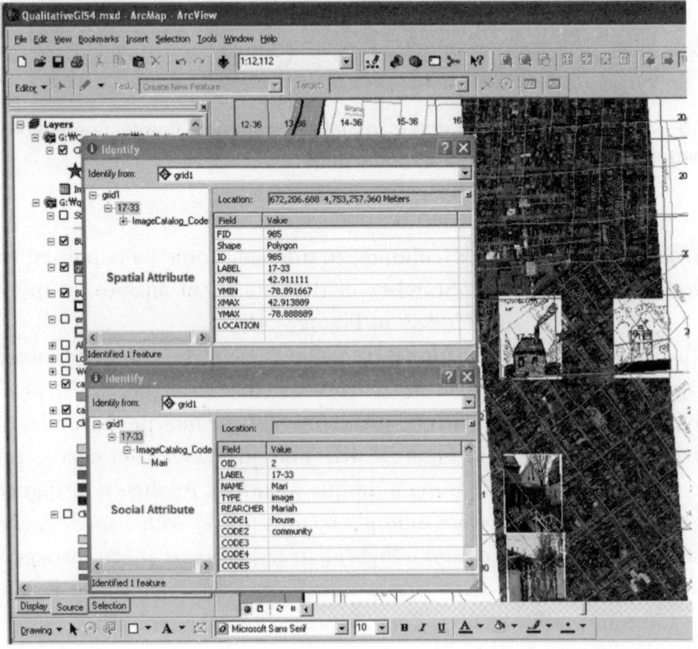

Figure 7.5 Information search result in CAQ-GIS

on children's urban geographies. I had gathered multiple forms of data, including conventional forms of GIS-based data (such as census data or satellite images) as well as qualitative images, such as the neighborhood images shown in Figures 7.2 and 7.3. Other qualitative images included paintings created by participating children when I asked them to express their thoughts about the meanings of community in any medium. Figure 7.5 is a screenshot that shows some of the visual displays and tables that resulted when I stored all of these data in a GIS using an imagined grid and a hybrid relational database.

Using these structures, the data may be explored in many ways. If we want to know more about the sketch shown next to the star symbol on the right of Figure 7.5, we can use ArcGIS tools to click the image and access related information. The two 'Identify Results' windows in Figure 7.5 show the results of such an action: retrieval of the tabular information contained in the spatial and qualitative data tables of the hybrid relational database. The spatial data table, which is the upper 'Identify Results' window, shows the georeferenced information that defines the location of the qualitative image (the sketch) stored in this cell of the imagined grid. The fields XMIN, YMIN, XMAX, and YMAX contain the latitude and longitude information of that grid cell. The qualitative data table, shown in the lower 'Identify Results' window, stores multiple types of qualitative data, as well as information such as type of data, researcher's name, label, and description of data. The association of multiple artifacts or attributes in the qualitative data table with each record in the spatial data table is made possible through the relational join function in ArcGIS.

Together, the imagined grid and the hybrid relational database enable multiple forms of qualitative data to be included, enabling a user to explore and visualize these data in a GIS. By including the original qualitative data directly, rather than transforming it into quantitative forms, the full contextual richness of these data is preserved. Georeferencing and storage of qualitative data in GIS data structures addresses many of the limits of hyperlinking strategies. Further, the possibility of storing interpretive codes in the hybrid relational database lays the foundation for using qualitative techniques to analyze these data. But simply storing data and codes in a GIS is somewhat limited in terms of the qualitative analytical operations that might be performed in the GIS itself. In the following section, I show a way of linking my GIS structures with computer-aided qualitative analysis software, so that the many analytical functions of such software might be brought to bear upon the data stored in the imagined grid and the hybrid relational database.

LINKING GIS AND COMPUTER-AIDED QUALITATIVE ANALYSIS SOFTWARE: CAQ-GIS

My efforts to support qualitative data analysis techniques in a GIS environment focus specifically on inductive techniques such as grounded theory. Grounded theory is an inductive analysis technique that enables the generation of theoretical propositions from qualitative data. In grounded theory, researchers are not testing preconceived theoretical propositions but rather are performing iterative inductive analysis that will allow the theory to *emerge* from the data (Strauss and Corbin, 1998: 12–14). Grounded theory is especially useful in dealing with the complexity and nuance of qualitative data because it emphasizes exploring multiple possibilities or interpretations; using multiple media to represent and explore data, including visual aids such as diagrams, memos, and conceptual 'networks'; and using non-linear forms of thinking such as iterative analysis. Overall, grounded theory is a technique that conceives of analysis as a dynamic interplay between researchers and their data (Charmaz, 2000; Chiovitti and Piran, 2003; Strauss and Corbin, 1997; 1998).

The heart of grounded theory is the process of coding (Lonkila, 1995). Coding is a way of evaluating and organizing data to understand meanings in the text in a systematic way (Cope, 2003; 2005; Miles and Huberman, 1994). Codes are labels for assigning units of meaning to the particular qualitative data artifacts, such as a qualitative image or a textual narrative. Codes may come from the researcher's conceptual framework, or from words and themes contained in the data themselves. The process of creating and assigning codes is a kind of analysis, directed at finding meanings and relationship within the data by differentiating or combining data (Miles and Huberman, 1994; Miles and Weitzman, 1996). Coding is an important part of analysis because it identifies key ideas and repeated themes. In the process, it is a way of filtering data, creating a manageable dataset, and dealing with a huge amount of information without discarding the detail and importance of the original data (Strauss and Corbin, 1998). Within a grounded theory framework, the process of coding is part of building theory, not testing it, though of course codes and coding processes may be influenced by existing theoretical frameworks (Burawoy, 1991). Researchers have long done grounded theory and coding without using computing technologies, and many still do so. But over the past 20 years, a number of software packages have been developed for helping researchers perform qualitative analysis more efficiently and effectively. CAQDAS packages are designed to assist the researcher in managing the analysis of qualitative data, and help them develop conceptual propositions from their qualitative data.

There are many different types of CAQDAS packages in use today, including word processors, text retrievers, text base managers, code-and-retrieve programs, code-based theory builders, and conceptual network builders (Kelle, 1995; Tesch, 1990; Weitzman and Miles, 1995). The search, highlighting, and commenting functions of ordinary word processing programs can be used for coding and grounded theory. Text retrievers do precisely what their name suggests: enable the researcher to search on particular words or phrases, retrieving these instances and often providing a count of their number. Text base managers operate similarly, but have more sophisticated content analysis functionalities for managing large databases (Lewins and Silver, 2006). Code-and-retrieve programs and code-based theory builders incorporate many of these same functions, but also allow researchers to code their data. These types of CAQDAS can also typically include visual forms of data (not only text) and provide visualization tools that researchers can use to create graphic visualizations of connections among themes or codes (Miles and Weitzman, 1996; Weitzman, 2000). The most important point about any of these CAQDAS types is that they do not automate or *do* qualitative analysis. As Weitzman argues, 'Software cannot pull good work out of a poor qualitative researcher, but it can in fact help competent researchers do more consistent and thorough research' (2000: 817). As he suggests, CAQDAS packages provide tools that researchers can use to facilitate their analysis, such as conceptual network tools or the ability to store and retain codes and associations in the data in digital form.

CAQDAS packages have not typically supported spatial data analysis directly, or the inclusion of spatial databases, though geographers and others are calling for such capabilities (Crang et al., 1997; Hoven, 2003). Further, several GIS researchers have proposed taking advantage of these strong data management and analysis functions of CAQDAS to support qualitative analysis in a GIS environment. Kwan (2002b) notes that GIS and CAQDAS both support visualization and query of visual forms of data or evidence, links to qualitative data such as photos and voice clips, and query tools based upon Boolean operations. Thus, she proposes, linking these two softwares could be one strategy for supporting qualitative analysis in a GIS environment. While Kwan (2002b) does not implement this proposition in practice, Matthews et al. (2005) have built an interface that tries to combine GIS and a CAQDAS package to support their 'geoethnography' technique. This use of GIS and CAQDAS together to support a very large collaborative ethnographic project is a tremendously important step forward. But it is limited in some ways because the GIS database and the ethnographic database in CAQDAS are still largely separate. In their system, the GIS is primarily used for visualization, with audio, video, image, and text files associated externally through hyperlinks. The ethnographic data analysis is also carried out separately, using the CAQDAS. My approach builds on these existing approaches to more strongly connect the two software systems for the purposes of analysis.

To use GIS with CAQDAS in an integrative way, in support of recursive iterative analysis of multiple forms of data, we need a way to connect the two systems. I propose using qualitative codes to accomplish this, using the codes as an anchor or bridge between GIS and CAQDAS. I refer to the resulting integrated system as a computer-aided qualitative GIS (CAQ-GIS). I choose to use codes to create this link for pragmatic and analytic reasons. Codes and coding are the essential elements of grounded theory, so we must support them in GIS and in CAQDAS if we are to build a truly integrated system for this sort of qualitative analysis. But at the practical level, codes and coding are already embedded in CAQDAS, and my hybrid relational database structure allows them to be integrated into the GIS database as well. This creates a critical link between CAQDAS and GIS. Thus, the codes created by a researcher as s/he analyzes qualitative data function as a 'bridge' that connects these two types of software.

I implemented this approach using my ArcGIS and a CAQDAS package called ATLAS.ti, an excellent code-based theory builder program. Its qualitative data storage and analysis tools are organized into several modules within the software. In ATLAS.ti, a project or 'hermeneutic unit' may comprise primary documents, smaller quotations, codes, and memos (researchers' notes that can be associated with primary documents, quotes, and so on). ATLAS.ti can help researchers to organize a variety of data types such as text, graphics, audio, and video data files. This capability of handling text-based data as well as non-text-based multimedia data is the essential strength that distinguishes this program from other CAQDAS programs and makes it ideal for my approach.

Figure 7.6 summarizes the system that I developed for linking GIS with CAQDAS. It outlines the structures that comprise CAQ-GIS in its entirety. The qualitative GIS,

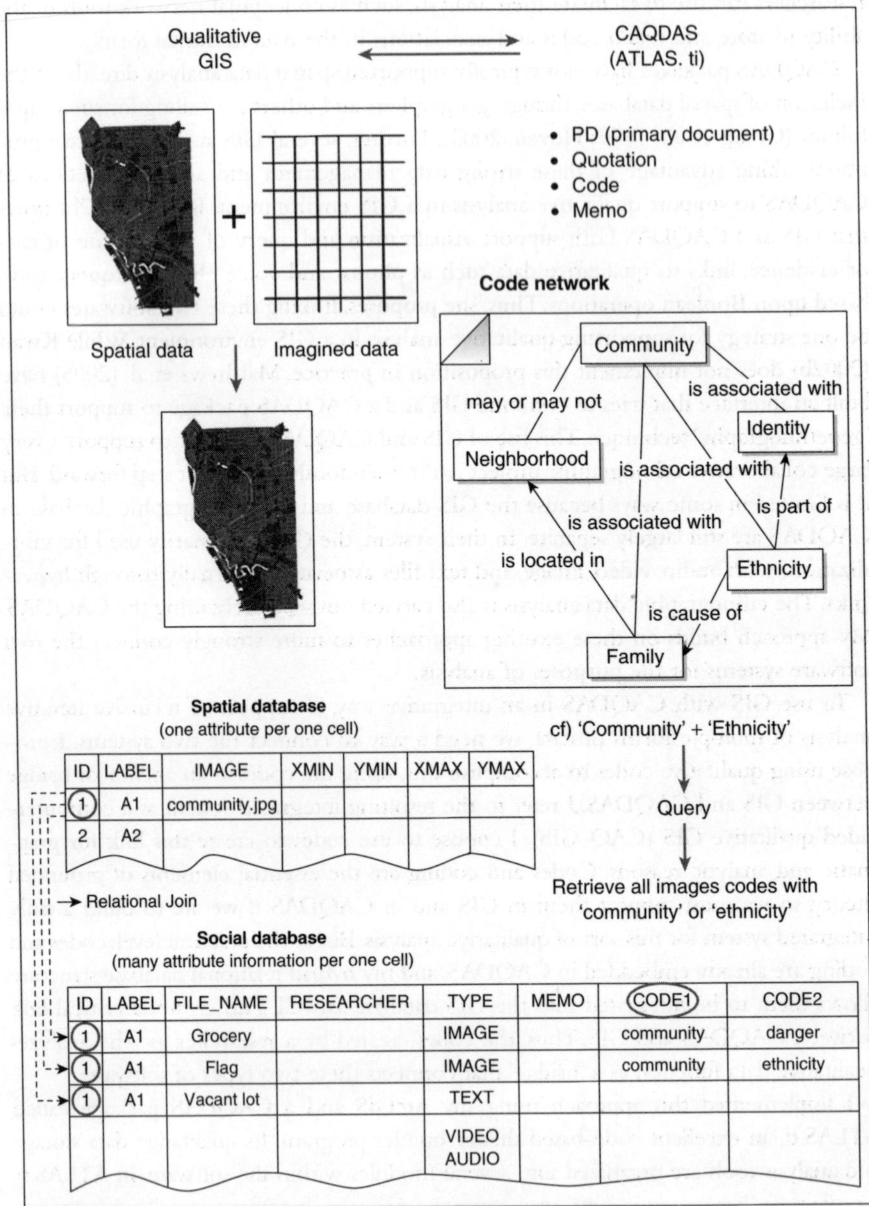

Figure 7.6 CAQ-GIS design

shown on the left side of the figure, consists of the imagined grid and the hybrid relational database and is used to store qualitative data and qualitative codes. The CAQDAS is shown on the right side of Figure 7.6. There I also show an example of the sort of conceptual networks that a researcher can create relating different codes (and the data to

Figure 7.7 ATLAS.ti search results for code 'community'

which they are assigned). In this code network diagram, codes related to neighborhood, family, ethnicity, community, and identity are shown, along with statements that conceptualize their possible relationships with one another. In my approach, these same codes would be stored in the hybrid relational database, attached to the qualitative data stored there. Thus, in setting up a CAQ-GIS, the researcher includes a field to store codes in the GIS database. Such fields are already present in the data structures of a CAQDAS. As researchers code their data, they populate these fields with codes based upon their analysis of specific geographic entities that are represented in the GIS.

This integration of GIS and CAQDAS makes it possible to retrieve all the qualitative data stored in ATLAS.ti that is linked with particular codes, and to do the same within the GIS software. Figure 7.5 shows the result of such a search and retrieval in the hybrid relational database. From within ArcGIS, the image that was clicked had two codes linked to it in the hybrid relational database: 'code 1: house' and 'code 2: community'. A search on the same code in ATLAS.ti will retrieve all data related to that code that are stored in the ATLAS.ti database, as shown for the code 'community' in Figure 7.7. In this example, this action has retrieved several different types of qualitative data, including the video clip that is shown in Figure 7.7.

These connections between CAQDAS and GIS that are established through qualitative codes make it possible to work simultaneously with the two systems in a coordinated way. Figures 7.8 and 7.9 show the results of such a parallel search and retrieval process, based upon the codes 'community' and 'ethnicity'. For example, in my research with children about their perception of the meanings of community, I stored qualitative

Figure 7.8 Search results for codes 'community' and 'ethnicity' in CAQDAS

data from the research in an ATLAS.ti database. I analyzed them with ATLAS.ti's network manager function, which is a very useful tool for developing and observing relationships among different codes or data. To explore possible associations between the meanings of community given by children and their ethnicity, I retrieved all data which had been coded as both 'community' and 'ethnicity' using ATLAS.ti's query tool. This action retrieved six data items: one text file, one movie file, and four images. The screenshot shown in Figure 7.8 displays some of these items as they were returned by the query in ATLAS.ti.

A parallel query executed in ArcGIS with these same codes will retrieve all qualitative data in the GIS database that carry the same code. Figure 7.9 shows the results of such a query when it is carried out in ArcGIS. The highlighted rows in Figure 7.9 show a list of qualitative data in a GIS hybrid database, which were codes with both 'community' and 'ethnicity'. Figure 7.9 shows these records in the data table that are returned by the query (names 'flag' and 'F3–2'), as well as the georeferenced neighborhood images in the map display that would also be selected as a result.

These parallel queries and data retrievals are made possible by the qualitative codes that bridge GIS and CAQDAS. These structures enable a researcher to move interactively between the two systems, recursively exploring and analyzing data that are stored in each. With this system, researchers have available to them the rich spatial analysis and visualization tools of GIS, as well as the conceptual network tools and other grounded theory tools that are available in CAQDAS. Thus, my CAQ-GIS can support robust mixed methods research that incorporates quantitative data and qualitative data; spatial data and non-spatial data; and many modes of analysis, including grounded theory, grounded visualization, and any spatial or quantitative analysis technique that can typically be carried out in a GIS.

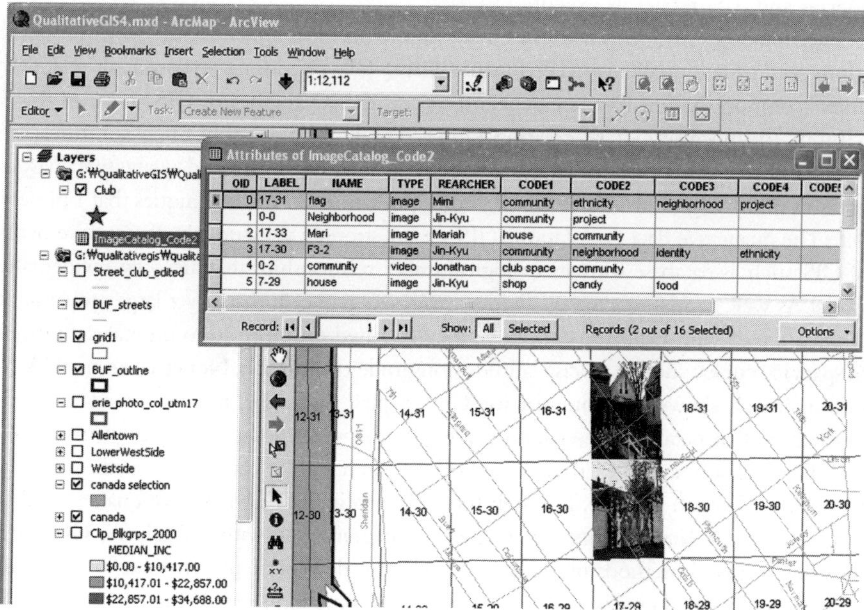

Figure 7.9 Search result for codes 'community' and 'ethnicity' in GIS

CONCLUSION

The system I have developed contributes to growing efforts by geographers and other researchers to support mixed methods research that integrates GIS with qualitative forms of data and analysis. Other existing approaches support qualitative data storage and analysis separately from GIS and its visualization capabilities. In contrast, my approach begins by georeferencing qualitative data so that they can be included in GIS data structures (and therefore in GIS-based visualization). Then, it enables researchers to carry out grounded-theory-based analysis that draws upon data stored in GIS data structures and in CAQDAS. This approach gives researchers access to the diverse analysis tools of GIS and CAQDAS in such a way that they may systematically explore multiple types of data using multiple modes of analysis.

In this chapter, I have described the key theoretical and technological concepts that inform my approach, and the software innovations I developed to implement it. First I created a way to include qualitative data directly into GIS data structures, using the imagined grid and the hybrid relational database. The imagined grid allows us to assign spatial information to qualitative data so that they can be located on a GIS-based map, and enables us to include these data in GIS. The hybrid relational database allows researchers to associate multiple qualitative data items with particular cells in the imagined grid, and gives us a way to add qualitative codes to these data. The second component of my approach is intended to facilitate inductive iterative analysis of these diverse data, through integrated use of GIS and CAQDAS. Qualitative codes, present in the data structures of both systems, can serve as a bridge connecting them. This enables systematic exploration of the data stored in each system through coordinated queries and data retrievals executed in each.

The resulting CAQ-GIS addresses many of the practical limits of other software-level efforts to facilitate qualitative GIS. Further, CAQ-GIS provides a powerful yet relatively simple solution that enables researchers to access the capabilities of GIS and CAQDAS without the necessity of re-creating either system from scratch. Such solutions are essential if researchers are to be able to adopt an integrated GIS and qualitative analysis system on any large scale. There are of course limitations to the techniques that I present here. The qualitative data stored in the GIS are still strongly structured by the requirements of GIS, such as database formats or logical query relationships that are only incidentally spatial. As well, georeferencing qualitative images to grid cells that cover large areas might confuse a user who expected data to be located based on more conventional definitions of spatial accuracy, such as precise latitude/longitude coordinates. Nonetheless, my CAQ-GIS approach shows the potential for GIS to play a valuable role in mixed methods research, not in parallel but separate practice with other methods, but in an integrated way. That is, more complex mixed methods research can be carried out with GIS than has previously been possible. Through the ability to include and analyze qualitative data *within* the GIS database, and by iteratively integrating it with CAQDAS, GIS itself becomes a mixed methods framework. I hope this development generates interest, discussion, and refinement by other researchers and practitioners seeking to break down methodological barriers and produce new approaches to support qualitative uses of GIS.

NOTES

1 The concepts of 'neighborhood' and 'community' have been debated at length in geography and elsewhere (Martin, 2002; 2003). Here I use the term 'neighborhood' and 'community' to mean those living spaces that residents identify as their neighborhood. Of course, there is a great deal of diversity in how residents identify these spaces and the meanings they attach to them, based upon their varying experiences, feelings, and social, cultural, and ethnic understandings.

2 The neighborhood image must first be clipped to match the size and shape of the grid cell.

REFERENCES

Aitken, S.C. and Michel, S.M. (1995) 'Who contrives the "real" in GIS? Geographic information, planning and critical theory', *Cartography and Geographic Information Systems*, 22 (1): 17–29.

Al-Kodmany, K. (2000) 'Extending geographic information systems (GIS) to meet neighborhood planning needs: Recent developments in the work of the University of Illinois at Chicago', *The URISA Journal*, 12 (3): 19–37.

Al-Kodmany, K. (2002) 'GIS and the artist: shaping the image of a neighborhood through participatory environmental design', in W. Craig, T. Harris and D. Weiner (eds), *Community Participation and Geographic Information Systems*. London: Taylor and Francis. pp. 321–9.

Barndt, M. (1998) 'Public participation GIS: barriers to implementation', *Cartography and Geographic Information Systems*, 25 (2): 105–12.

Burawoy, M. (1991) 'The extended case method', in M. Burawoy, A. Burton, A.A. Ferguson and K.J. Fox (eds), *Ethnography Unbound: Power and Resistance in the Modern Metropolis*. Berkeley, CA: University of California Press. pp. 271–90.

Charmaz, K. (2000) 'Grounded theory: objectivist and constructivist methods', in N.K. Denzin and Y.S. Lincoln (eds), *Handbook of Qualitative Research.* Thousand Oaks, CA: Sage. pp. 509–36.

Chiovitti, R.F. and Piran, N. (2003) 'Rigour and grounded theory research', *Journal of Advanced Psychology*, 44 (4): 427–35.

Chrisman, N. (1997) *Exploring Geographic Information Systems.* New York: Wiley.

Cieri, M. (2003) 'Between being and looking: queer tourism promotion and lesbian social space in Greater Philadelphia', *ACME: An International E-Journal for Critical Geographies*, 2 (2): 147–66.

Cope, M. (2003) 'Coding transcripts and diaries', in N.J. Clifford and G. Valentine (eds), *Key Methods in Geography*. London: Sage. pp. 445–60.

Cope, M. (2005) 'Coding qualitative data', in I. Hay (ed.), *Qualitative Research Methods in Human Geography*. Oxford: Oxford University Press. pp. 310–332.

Corbett, J.M. and Keller, C.P. (2005) 'An analytical framework to examine empowerment associated with participatory geographic information systems (PGIS)', *Cartographica*, 40 (4): 91–102.

Crampton, J. and Krygier, J.B. (2006) 'An introduction to critical cartography', *ACME: An International E-Journal for Critical Geographies*, 4 (1): 11–33.

Crang, M.A., Hudson, A.C., Reimer, S.M. and Hinchliff, S.J. (1997) 'Software for qualitative research. 1: Prospectus and overview', *Environment and Planning A*, 29: 771–87.

Dennis, S.F. (2006) 'Prospects for qualitative GIS at the intersection of youth development and participatory urban planning', *Environment and Planning A*, 38 (11): 2039–54.

Dorling, D. (1998) 'Human cartography: when it is good to map', *Environment and Planning A*, 30: 277–88.

Elwood, S. (2006) 'Critical issues in participatory GIS: deconstructions, reconstructions, and new research directions', *Transactions in GIS*, 10 (5): 693–708.

Elwood, S. and Leitner, H. (1998) 'GIS and community-based planning: exploring the diversity of neighborhood perspectives and needs', *Cartography and Geographic Information Systems*, 25 (2): 77–88.

Fiedler, R., Schuurman, N. and Hyndman, J. (2006) 'Improving census-based socioeconomic GIS for public policy: recent immigrants, spatially concentrated poverty and housing need in Vancouver', *ACME: An International E-Journal for Critical Geographies*, 4 (1): 145–71.

Flavelle, A. (1995) 'Community-based mapping in Southeast Asia', *Cultural Survival*, 18 (4): 72–3.

Ghose, R. (2001) 'Use of information technology for community empowerment: transforming geographic information systems into community information systems', *Transactions in GIS*, 5 (2): 141–63.

Ghose, R. and Huxhold, W.E. (2001) 'Role of local contextual factors in building public participation GIS: the Milwaukee experience', *Cartography and Geographic Information Systems*, 28 (3): 195–208.

Gilbert, M.R. and Masucci, M. (2005) 'Research directions for information and communication technology and society in geography', *Geoforum*, 36: 277–9.

Gilbert, M.R. and Masucci, M. (2006) 'The implications of including women's daily lives in a feminist GIScience', *Transactions in GIS*, 10 (5): 751–62.

Harris, L. and Harrower, M. (2006) 'Introduction. Critical interventions and lingering concerns: critical cartography/GISci, social theory, and alternative possible future geographies', *ACME: An International E-Journal for Critical Geographies*, 4 (1): 1–11.

Harris, L.M. and Hazen, H.D. (2006) 'Power of maps: (counter) mapping for conservation', *ACME: An International E-Journal for Critical Geographies*, 4 (1): 99–130.

Harris, T. and Weiner, D. (1998) 'Empowerment, marginalization, and community-oriented GIS', *Cartography and Geographic Information Systems*, 25 (2): 67–76.

Hoven, B.V. (2003) 'Using CAQDAS in qualitative research', in N. Clifford and G. Valentine (eds), *Key Methods in Geography*. London: Sage. pp. 461–76.

Johnson, J.T., Louis, R.P. and Paramono, A.H. (2006) 'Facing the future: encouraging critical cartographic literacies in indigenous communities', *ACME: An International E-Journal for Critical Geographies*, 4 (1): 80–98.

Kanarinka (2006) 'Art-machines, body-ovens and map-recipes: entries for a psychogeographic dictionary', *Cartographic Perspectives*, 53: 24–40.

Kelle, U. (ed.) (1995) *Computer-Aided Qualitative Data Analysis: Theory, Methods and Practice.* London: Sage.

Knigge, L. and Cope, M. (2006) 'Grounded visualization: integrating the analysis of qualitative and quantitative data through grounded theory and visualization', *Environment and Planning A*, 38 (11): 2021–37.

Krygier, J.B. (1999) 'Cartographic multimedia and praxis in human geography and the social sciences', in W. Cartwright, M.P. Peterson and G. Gartner (eds), *Multimedia Cartography*. Berlin: Springer. pp. 245–55.

Krygier, J.B. (2002) 'A praxis of public participation GIS and visualization', in W.J. Craig, T.M. Harris and D. Weiner (eds), *Community Participation and Geographic Information Systems*. London: Taylor and Francis. pp. 330–45.

Krygier, J.B. (2006) 'Jake Barton's performance maps: an essay', *Cartographic Perspectives*, 53: 41–50.

Kwan, M.P. (2002a) 'Feminist visualisation: re-envisioning GIS as a method in feminist geographic research', *Annals of the Association of American Geographers*, 92 (4): 645–61.

Kwan, M.P. (2002b) 'Is GIS for women? Reflections on the critical discourse in the 1990s', *Gender, Place and Culture*, 9 (3): 271–9.

Kwan, M.P. and Knigge, L. (2006) 'Guest editorial. Doing qualitative research using GIS: an oxymoronic endeavor?', *Environment and Planning A*, 38 (11): 1999–2002.

Lewins, A. and Silver, C. (2006) 'Choosing a CAQDAS package', *CAQDAS Networking Project*. http://caqdas.soc.surrey.ac.uk/index.htm.

Lonkila, M. (1995) 'Grounded theory as an emerging paradigm for computer-assisted qualitative data analysis', in U. Kelle (ed.), *Computer-Aided Qualitative Data Analysis*. London: Sage. pp. 41–51.

Martin, D.G. (2002) 'Constructing the "neighborhood sphere": gender and community organizing', *Gender, Place and Culture*, 9 (4): 333–50.

Martin, D.G. (2003) 'Enacting neighborhood', *Urban Geography*, 24 (5): 361–85.

Matthews, S.A., Detwiler, J.E. and Burton, L.M. (2005) 'Geo-ethnography: coupling geographic information analysis techniques with ethnographic methods in urban research', *Cartographica*, 40 (4): 75–90.

McLafferty, S.L. (2002) 'Mapping women's Worlds: knowledge, power and the bounds of GIS', *Gender, Place and Culture*, 9 (3): 263–9.

McLafferty, S.L. (2005a) 'Geographic information and women's empowerment: a breast cancer example', in L. Nelson and J. Seager (eds), *Companion to Feminist Geography*. Malden, MA: Blackwell. pp. 486–95.

McLafferty, S.L. (2005b) 'Women and GIS: geospatial technologies and feminist geographies', *Cartographica*, 40 (4): 37–45.

Miles, M.B. and Huberman, A.M. (eds) (1994) *An Expanded Sourcebook: Qualitative Data Analysis*. London: Sage.

Miles, M.B. and Weitzman, E.A. (1996) 'The state of qualitative data analysis software: what do we need?', *Current Sociology*, 44: 206–24.

Pavlovskaya, M.E. (2002) 'Mapping urban change and changing GIS: other views of economic restructuring', *Gender, Place and Culture*, 9 (3): 281–9.

Pavlovskaya, M.E. (2006) 'Theorizing with GIS: a tool for critical geographies?', *Environment and Planning A*, 38 (11): 2003–20.

Rose, G. (2001) *Visual Methodologies: An Introduction to the Interpretation of Visual Materials*. London: Sage.

Schuurman, N. (2000) 'Critical GIS: theorizing an emerging science'. PhD dissertation, University of British Columbia, Vancouver.

Schuurman, N. (2004) *GIS: A Short Introduction*. London: Blackwell.

Schuurman, N. and Pratt, G. (2002) 'Care of the subject: feminism and critiques of GIS', *Gender, Place and Culture*, 9 (3): 291–9.

Sheppard, E.S. (1995) 'GIS and society: towards a research agenda', *Cartography and Geographic Information Systems*, 22 (1): 5–16.

Sheppard, E.S. (2001) 'Quantitative geography: representations, practices, and possibilities', *Environment and Planning D: Society and Space*, 19 (5): 535–54.

Sheppard, E.S. (2005) 'Knowledge production through critical GIS: genealogy and prospects', *Cartographica*, 40 (4): 5–21.

Strauss, A. and Corbin, J. (1997) *Grounded Theory in Practice*. London: Sage.

Strauss, A. and Corbin, J. (eds) (1998) *Basics of Qualitative Research: Techniques and Procedures for Developing Grounded Theory*. London: Sage.

Talen, E. (1999) 'Constructing neighborhoods from the bottom up: the case for resident-generated GIS', *Environment and Planning B: Planning and Design*, 26: 533-54.

Talen, E. (2000) 'Bottom-up GIS: a new tool for individual and group expression in participatory planning', *Journal of the American Planning Association*, 66 (3): 279–94.

Tesch, R. (1990) *Qualitative Research: Analysis Types and Software Tools*. Bristol, PA: Falmer.

Weiner, D. and Harris, T.M. (2003) 'Community-integrated GIS for land reform in South Africa', *The URISA Journal*, 15: 61–73.

Weitzman, E.A. (2000) 'Software and qualitative research', in N.K. Denzin and Y.S. Lincoln (eds), *Handbook of Qualitative Research*. London: Sage. pp. 803–20.

Weitzman, E.A. and Miles, M.B. (eds) (1995) *Computer Programs for Qualitative Data Analysis*. London: Sage.

Wood, D. (2006) 'Map art', *Cartographic Perspectives*, 53: 5–14.

CONCEPTUAL ENGAGEMENTS

8

INTO THE IMAGE AND BEYOND: AFFECTIVE VISUAL GEOGRAPHIES AND GISCIENCE

Stuart C. Aitken and James Craine

ABSTRACT

With postmodernism came the admonition that social scientists look more fully at the power of visual representations. Our world, it was suggested, is flooded by images and spectacles to the extent that it is superficially and depthlessly contrived. Within this context, some geographers began writing about the power of GISystems to create spectacular map outputs. Often, these outputs were used to sway recalcitrant policy makers on crucial geopolitical topics. The power of representation was an important focus of discussion. Today, two things have changed. First, we have gone into the image. GISystems are now augmented by GIScience and sophisticated geovisualization techniques. These visual technologies probe and explore the depths of data, creating new ways to think spatially. Second, we have gone beyond the image. Our understanding of the power of representation is now augmented with non-representational ways of knowing. This perspective goes beyond images to look at their affects. To look at affects spatially is to presuppose how emotional geographies change the world. With this chapter, we bring together these two, often disparate, knowledges to argue for a broader understanding of the affective power of visual geographies as part of our disciplinary systems of technology.

> In 1654, Madelaine de Scudéry published a map of her own design to accompany the novel *Clélie*. Her *Carte du pays de Tendre* – a map of the land of tenderness – pictures a varied terrain comprised of land, sea, river, and lake and includes, along with some trees, a few bridges and a number of towns. The map, produced by the female character of the novel to show the way of the 'countries of tenderness', embodies a narrative voyage … In this way, the *Carte de Tendre* makes a world of affects visible to us … Emotion materializes as a moving topography. (Bruno, 2002: 2)

There has been quite a lot of talk recently about emotions and other aspects of geographic practice that go beyond representational ways of knowing. Maps are, at base, representations and yet it is not an overstatement to suggest that when they represent space well they also draw us in imaginatively and emotionally. Many of us became geographers because we were at some time transfixed by this power of maps; as our minds wander around contours, our imaginations wander around those slopes. There is power to maps that is about what they are *and* what they do. Emotional geographies offer a theoretical basis for taking a look at this entanglement between visual objects and their affects. One root of the discussion about emotional geographies comes from poststructural theories that go beyond texts and other representations (Thrift, 2004; Wylie, 2006), while another is spurred by feminists concerns for elaborating theories of affect (Anderson and Smith, 2001; Davidson et al., 2007). In the epigraph with which we start this essay, Giuliana Bruno points to a seventeenth-century *cartography of tenderness* that countered the imperial cartographies of the day. We think that the notion of a tender mapping is hugely appropriate to moving into and beyond the imperial cartographies of today, and we believe that emotional geographies help us get to that place.

With this chapter, we argue that although geovisualization techniques open up new ways of exploring spatial data, these are more often than not still bound by mechanistic logics of representation. Alternatively, it may be useful to situate GIScience as a whole within its emotional and affective contexts. We consider the implications of affective visual geographies before raising recent examples that are specifically contextualized, like Bruno's example, within cartographies of affect. We then point to the implications of this move for affecting change in the ways GIScience is practiced and realized. We begin by setting up our arguments for affective geovisualizations.

THE NEED FOR AFFECTIVE GEOVISUALIZATIONS

It seems to us that spatial data visualized through geographic information systems – in cartographic form or otherwise – can be joyless and overcalculated, with a tendency for the program to overwhelm the content. Geovisualization is a subfield of GIScience that develops techniques and tools designed to interactively 'visualize' spatial phenomena. It uses exploratory techniques to help visualize spatial data in ways that elaborate patterns and processes. We contend that even the best geovisualized data are often more interesting to *think* about than to *experience*, more interesting to *create* than to *comprehend*; they are most often *less* the creation of a searching soul and *more* the product of a highly computer-literate mind (Aitken and Craine, 2006). And yet, as neuroscientist Antonio Damasio (1994) has shown, emotions are a huge part of, and are not separate from, our intellectual reasoning. From this union of emotional experience and data representation, we bring forth our 'affective geovisualization': a methodology that attempts to elaborate how humans interact with GIS-based digital virtualized environments and datasets.

When we coined the term 'affective geovisualizations' in the title of a 2006 article in *Directions Magazine* (Aitken and Craine, 2006) our intent was to suggest a link between emotive work in geographic studies of media, particularly film, and some of the cognitive and behavioral work in geovisualization. Our argument was that these two fields of study tend to talk past each other and it was time to more closely look at the ways that they might inform each other. Given the early contributions to geovisualization of semiotics and other structured ways of knowing how representations work on viewers (MacEachren, 1994), it seemed reasonable to suggest that non-representational theory, with its focus on emotions, could well play an important part in theorizing the current contexts of geovisualization. With this chapter, we want to move these ideas forward a little more with a discussion of the ways visual geographies coalesce with, in, and through GIScience. We want to suggest a way to proceed and, perhaps also, a way out of the dilemma of the 'infinite vision', the 'promising vision from everywhere and nowhere equally and fully' applied by Donna Haraway (1991: 189, 191) when she famously christened maps and other representations of 'Science' as 'an illusion, a god-trick'.

Here, we take Haraway's work a step further. We place the construction of the GIS dataset within the spaces of affective technologies in an attempt to understand the social power associated with these spaces. Haraway (1989; 1991), for example, argues how particular forms of visuality produce specific images of social difference and also how institutions, such as capitalism, mobilize certain forms of visuality to see, and to order, the world. Haraway urges us to embrace new technologies as we seek further insights into the social relations contained therein: 'taking responsibility for the social relations of science and technology means refusing an anti-science metaphysics, a demonology of technology, and so means embracing the skillful task of reconstructing the boundaries of daily life' (1991: 181). We argue that the vast amounts of digital data contained within GIS representations of visuality (which, in itself, *could* be considered a scientific discourse and is certainly a technological if not a social movement) are ideal sites for the study of social relations. While Wasko (1995) points out that new digital formats give unprecedented access to individuals, the reverse is also true: individuals now have unprecedented access to vast amounts of geographic information, including information not available to original analog audiences.

It is our concern that GIS and its technologies and practices have become so ubiquitous that they are, for all intents and purposes, invisible or *indiscernible*. Our purpose is to make these technologies and practices both visible and comprehensible through this new way of engaging GIS. Geographic representations such as maps are produced and consumed in historically specific and carefully constructed ways and many factors combine to frame the ways in which meaning is generated. Thus, these representations cannot be engaged in isolation from, but rather must be linked in multiple and complex ways to, other forms of material evidence. We are concerned with GIS not merely as a digital reflection of the 'real' world or as the intentions of its users, but as discrete moments in the production and circulation of meaning. Not only is GIS a technology of image making but, more importantly, it is a technology of information transfer and knowledge production, and thus functions as an act of communication. It is, therefore, a chain of practices and processes through which geographical information is gathered, geographical facts are ordered, and our imaginative geographies are constructed.

The title of our chapter comes from Kevin Robins's 1996 book in which he suggests that with postmodernism social scientists needed to look more fully at the power of visual representations. Our world, he argued, is flooded by images and spectacles to the extent that it is superficially and depthlessly contrived. The power of representation was an important focus of his discussion. Today, two things have changed. First, we have gone *into* the image. GISystems are now augmented by GIScience and sophisticated geovisualization techniques. These visual technologies probe and explore the depths of data, creating new ways to think spatially. Second, we have gone *beyond* the image. Our understanding of the power of representation is now augmented with non-representational ways of knowing. This perspective goes beyond images by looking at the affective properties of the image. It is about how emotional geographies change the world. To go into and through the image, we aver, is to argue for a broader understanding of the affective power of visual geographies as part of our disciplinary systems of technology.

TRADITIONAL VISUAL GEOGRAPHIES AND THEIR CRITIQUE

Geography is a visual discipline. In his presidential address to the Association of American Geographers, Carl Sauer admonished that geography 'is first of all knowledge gained by observation, that one orders by reflection and re-inspecting the things that one is looking at ... often interrupted by leisurely halts to sit at vantage points and stop at question marks' (1956: 296). This penchant for leisurely observation from vantage points was critiqued by feminist geographers, most notably Gillian Rose, who argued against the patriarchal nature of this kind of desire through seeing: 'The inherent fears in geography's visual pleasures, its suspicion in its pleasure, produce its persistent refusal to problematize its pleasure – geographers are invisible to themselves' (1993: 107). As part of a project that pushed against what was seen as a larger patriarchal project in science, feminists evoked Donna Haraway's (1991: 189) simian/cyborg manifesto, which declared the need for a usable, but not innocent, doctrine that recognized the possibility of sustained, rational, objective enquiry residing solely within a politics and epistemology of partial and situated perspectives.

A series of powerful critiques in the 1990s problematized geography's traditional visual focus not just for its patriarchy but also on a host of other power relations that are imbedded in visual phenomena such as maps, landscapes, paintings, advertisements, photographs, and movies (Crang et al., 1999; Cresswell and Dixon, 2002; Fyfe, 1998; Robins, 1996). These critiques targeted both the sciences and the arts. As Rose points out, 'Science' may strive for disembodied mastery, but with the ideological notion that 'Art' is the ultimate form of human expression comes the problematic assumption that 'its pleasure is assumed to be untainted by the specificity of social relations' (1993: 99).

Nonetheless, the appeal of visual geography still resides with the ways it uses all types of visual images (maps, paintings, photographs, animations, movies) to highlight the physical world as a metaphorical space for the portrayal of spatial and social relationships. This, we argue, is a good start but is at best a naive understanding of the ways images stand in for something else. It glosses over how little we know about the power of the visual. We do know that images are powerful and that they are used for good and ill, but we do not really know the extent of their influence, nor do we know with any certainty what aspects of images are seen, or how we connect with them, and who connects in what ways. In addition, despite 50 years of analysis in semiotics – a science that focuses specifically on linguistic systems and signs – no one has come up with a precise relationship between images and language structures. Nor do we understand well the emotive and non-representational aspects of visual representations. In the face of this uncertainty, it is intriguing to look at the ways that images are used in geographic information sciences.

GEOVISUALIZATIONS

> [V]isualization is foremost an act of cognition, a human ability to develop mental representations that allow geographers to identify patterns. (MacEachren, 1992: 101)

> [A]ffectively imbued thinking is always already under way by the time consciousness intervenes to pull it in this or that direction. (Connolly, 2002: 94)

Computer-generated multi-dimensional representations do not initially explain phenomena or events, but rather enable researchers to see things that might not be immediately obvious or visible. As part of this trend in science, more sophisticated technological representations of spaces, places, patterns, and processes through geographical information science highlight a visual acuity that takes the viewer well beyond simple cartographic displays. In contrast to cartographic visualization, geovisualization's aim is not normally to communicate knowns, but rather to explore unknowns (MacEachren, 1994; MacEachren et al., 2004). As the first epigraph to this section suggests, geovisualization is primarily about developing mental representations. It emerged as a field in the 1990s as a combination of scientific visualization, information visualization, and exploratory georeferenced data analysis (Buttenfield et al., 2000). Characterizing the key components of representation is an important part of geovisualization. For example, Skupin and Fabrikant (2003) provide what they call an itinerary of geovisualized representations of non-geographic information, and Dykes (1997) uses dynamic graphics to elaborate cartographic and statistical representations. Acevedo and Masuoka (1997) and Fabrikant (2005) use animations to represent spatial data with the purpose of exploring the scientific usefulness of different representational qualities.

Gahegan and Pike take on the context of situated knowledge representations, pointing out that there is now significant progress 'towards better representational models in computer and information science, which help to bring to light many aspects of meaning and epistemology that currently lurk in the backgrounds of many GIS applications' (2006: 728). They suggest some important ways forward in using computational representations to highlight the situated knowledge of what often resides in the 'volatile memory of the analyst(s)' as well as giving 'voice to different individuals and communities, specifically by representing a wealth of different perspectives that might be taken on geospatial information' (2006: 730). Influenced by Pierce's classic semiotic theory and Whitehead's foundations of knowledge representation, Gahegan and Pike go on to argue that 'some aspects of the situations surrounding creation and use of resources can be harvested, remembered, mined, visualized and applied to help communicate some … missing aspects of meaning, and to complement the more objective, top-down knowledge that might be provided by computational ontologies' (2006: 730). With this work, there is still an overriding focus on the representation and cognition where cognitive models are used to cover the ways users conceptualize, learn, and understand the data presented to them.

This focus misses an important new stream of work in neurology that highlights the emotional aspects of being in the world. As the second epigraph at the beginning of this section suggests, to discount the affective components of thought is perhaps to miss a foundational aspect of cognition. Challenging traditional views about the connections between emotions and rationality, neuroscientist Antonio Damasio (1994) suggested that Descartes's error in his famous proclamation 'I think, therefore I am' was to steer science away from emotions as the primary source of humanness. More recently, V.S. Ramachandran (2008), whose early work focused on visual psychophysics, suggests that what he calls 'mirror neurons' are the basis of human empathy and creativity in that they enable humans and primates (and some birds) to recognize and imitate the behaviors of others. In spite of the popularity of this field, to date no plausible neural or computational models have been developed to describe how mirror neuron activity supports cognitive functions, such as imitation, let alone emotional predispositions.

Using GIS, anyone can visually construct – or deconstruct – a spatial reality, and the result is a profound experiential and epistemological shift undergone by an increasingly digital culture. Referentiality and legitimation are finished with construction of space: there is no longer an unproblematic and empirically verifiable 'real' to refer *to*. What is left is esthetic and affective. We argue then that a process of affective geovisualization may be derived through an ontology of its visual space: the images of a GIS that constitutes itself as image. As Barthes (1971) noted, the photograph is an object, a record of human vision and presence, a frozen moment of the past, while we would argue that GIS enacts a present-time *experience* of physical, bodily spatial reality.

So again we go through and beyond the image. Geography has long used remotely sensed visual information to expand its topographies, first through analog aerial photography and then into digital satellite imaging with a resolution that increases with astounding rapidity; thus, the image is indeed firmly placed within the discipline. There are, however, crucial differences between the visual techniques already in use and what we propose through our application of affective geovisualization. The digital, processed graphics of geovisualized technology produce a constant mutation very much apart from the metamorphoses of human time and experience. The weakened representational function of these malleable images produces little sense of permanence, history, or bodily investment at all.

The advent of digital domains of geographic knowledge has given a new, and ontologically different, life to geography. Through affective geovisualization we can now 'see' what human senses cannot: the movement of geographic information through virtual environments. We now have a new realm of images with an extended, but still valid, reconstruction of the 'real'. GIS and its digital spatialities offer previously undiscovered sites of geographic exploration: the modeling of three-dimensional, virtualized objects is another means of envisioning geographic space.

Current geovisualization discourse has not quite moved out of the technological implications of GIS and into more human-based emotive and affective spatialities.

In the balance of this chapter we argue that one of the problems of geovisualization studies is this continued focus on computational and cognitive modeling that misses Damasio's earlier point that emotions are essential to rational thinking and normal social and spatial behavior. In addition, we are concerned that geo-representations are generally understood as given, or at best as spatial metaphors standing in for social and spatial processes and thereby missing the nuances of constant mutations in meanings, interpretations, and subjectivities. We begin with a look at the power of representations as they move towards spectacle before turning to affect as a way to encounter the powerful non-representational aspects of images.

REPRESENTATIONAL DEPTHLESSNESS

Despite the critique of traditional notions of the visual, it is worth reiterating that representations give back to society an image of itself and they are, as such, one of the most important and oldest aspects of theatre and art, as well as science. From the perspective of scientific geovisualization (as well as art and theater), representations work because they reinforce a set of societal structures that help to make sense of contexts that are otherwise chaotic and seemingly random, and they help us define ourselves in relation to those contexts. Postmodern theorists in the 1980s and 1990s argued that the political force of representation was heightened with and through spectacle. Given that visual representations are now an important part of scientific exploration begs questions about the nature of representation and the ways it may be transformed into spectacle.

In the 1980s, postmodern scholars argued that there is a planned depthlessness – what some called an amnesia – in the work of images. To understand this kind of work (which is not necessarily the product of one person or a team of engineers: indeed, it is often something that foments outside the planning process and has important historical precedents), it is important to focus on the ways images permeate society in the form of spectacles. While images are often organized in the interests of assuring narrative significance, they also develop as something fascinating in and of themselves and by so doing they become unshackled from their social and cultural roots. This source of visual pleasure is predicated upon the creation of spectacle. At a number of different levels – and precisely because of this depthlessness, this amnesia – visual spectacles create emotional fervor that may become a deeply insidious political tool. An example may help here.

Throughout 2006 *The San Diego Union–Tribune* published a number of maps and renderings of San Diego's Embarcadero to highlight the influence of the Manchester Financial Group on the redevelopment of property owned by the US Navy. These images moved towards spectacle when combined with text that elaborated the plans of developer Douglas Manchester to build 2.9 million square feet of offices, condominium-hotels, and retail stores on the 14.7 acre site along Harbor Drive. The spectacle of 40-storey buildings along San Diego's waterfront was very

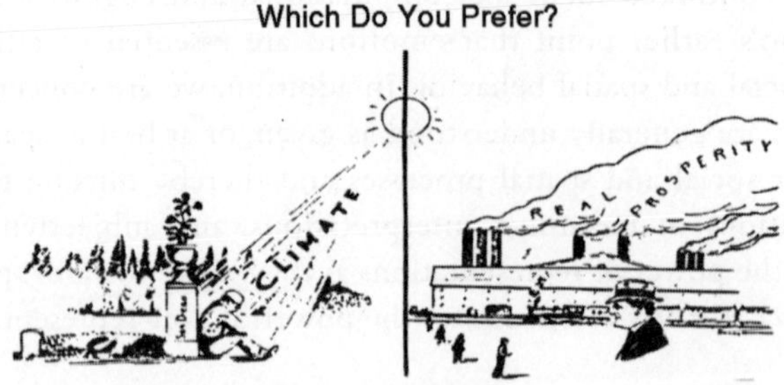

Figure 8.1 The spectacle of San Diego's 1917 mayoral campaign literature. The significance of the ways these representations come together to suggest spectacle is an important (e)motivator of public reaction

much a part of – indeed it may be argued that they were wholly the propagator of – public engagement with this project.

Roger Showley's (2006) article 'On the Waterfront: Battle Lines Drawn' in the *Union–Tribune* weighed into the context of this spectacle with a title – taken from Kazan's classic 1954 movie starring Marlon Brando – that evokes a battle between working people and larger institutions. Strategically, as part of the *Union–Tribune's* occasional 'Smokestack and Geraniums' series, Showley's work in turn invoked a spectacle of its own. The cornerstone of San Diego's 1917 mayoral campaign (Figure 8.1) pitted quick-fix industrial forces (smokestacks) against advocates of long-term planning (geraniums). Showley deftly positions the waterfront debate as the final chapter in a 100-year-old war on the future of San Diego's harbor. The first, he notes, was in 1803 and resulted in a standoff between American sailors and Spanish soldiers. The current debate involved the military to the extent that the Navy is uneasy with the idea of San Diego citizens telling it what to do with its property.

The late Berkeley political scientist Michael Rogin argued that 'spectacle is the cultural form for amnesiac representation, for spectacular displays are superficial and sensately intensified, short lived and repeatable … Spectacles colonize everyday life … and thereby turn domestic citizens into imperial subjects' (1990: 106). Pushing this reasoning further forward brings us to the famous admonition of Debord (1983/2000) that defines spectacle as the transformation of desire and fantasy into the reality of the commodity.

Robins (1996) explains how this may happen, arguing that if spectacle is created through certain kinds of images (e.g. high-rise towers, reality television, political speeches) it anesthetizes our sensibilities to the extent that our moral experiences are forgotten in favor of thrill and immediate 'presence'. The maps published in the *Union–Tribune* created a foundation in reality for the rendering, which is an artist's creation, that was then combined with text to stimulate a reaction to a potential presence. Fragmented and thereby diminished, the material realities of social and spatial relations are confused with the immediate presence and thrill of spectacle. Evoking Haraway's simian/cyborg manifesto, Robins (1996: 86) goes on to suggest that visual technologies herald a problematic utopianism precisely because 'no place' is 'every place'.

In this sense, rather than creating exploratory images for new data analysis – and we don't think this is too much of a stretch–geovisual technologies may create utopian visual spaces in the same sense that the images in the *Union–Tribune* created the spectacle of a future San Deigo waterfront. We'll have more to say about utopia and geovisualization in a moment; but it is important to note at this time that Robins (1996: 17) questions the nature of desires that constantly seek the 'other' of any space or place that cannot be satisfied by real spaces or places. Debord's logic suggests that these desires are easily commodified in global digital worlds. This so-called escape from reality is pernicious when combined with moral and memory decay. Geospatial images and information technology combine to form a utopian myth of better, digital worlds, or worlds that can be made better by digital surveillance and manipulation.

In looking for the origins of this development, Robins comes up with a chilling narrative that sandwiches consumerist needs for reality television with the pervasiveness of war technologies. The brutal realities of the contemporary warmongering West are diminished and our sensibilities to the cruelty that is part of that brutality are anesthetized by images that push back representations in favor of presence. In other words, we deny the real and shocking experiences of living in a fearsome and violent world with the presentation of spectacular images that do not require reason, analysis, or reflection (Robins, 1996: 121). And, ultimately, emotions that may push us to political acts are also anesthetized. In the same way that visual technologies create 'a world whose reality has been progressively screened out' (1996: 13), they also create a desensitization of responsibility and care.

TENDER MAPPINGS

Robins evokes a narrative that brings together desires that are fulfilled by consumerism and the denial of a brutal and violent world. He suggests, like Debord before him, that the pervasiveness of spectacle in contemporary society creates amnesia and anesthesia as balms against responsibility and care. And yet there is another narrative – another 'push in the world' to use Nigel Thrift's (2004: 186) term – that is worth considering.

In an astonishing and extensive reworking of cartography away from its imperial roots, Giuliana Bruno (2002) engages the contexts of mapping in architecture, travel, design, housing, planning, and film. She takes the history of mapping and contextualizes it in the arts, in desire, and in tenderness. Bruno calls her reworking of cartographic themes against prevailing imperial hegemony a 'sentimental geography' (2002: xi). As a starting point of her *Atlas of Emotion*, a work that moves in, between, and through seventeenth-century cartographies to twentieth-century films, Bruno evokes Madelaine de Scudéry's map that accompanies the novel *Clélie* (1654). Scudéry's *Carte du pays de Tendre* (Figure 8.2) is a celebrated allegory for the female association of desire with space, and an exemplar of the ways that cartography is inextricably linked with the shaping of female subjectivity (Benjamin, 1986). It highlights important passages and mobilities away from lakes of indifference, dangerous seas, and *terrae incognitae* to favorable villages and towns of tenderness, large hearts, reflection, sympathy, and so forth.

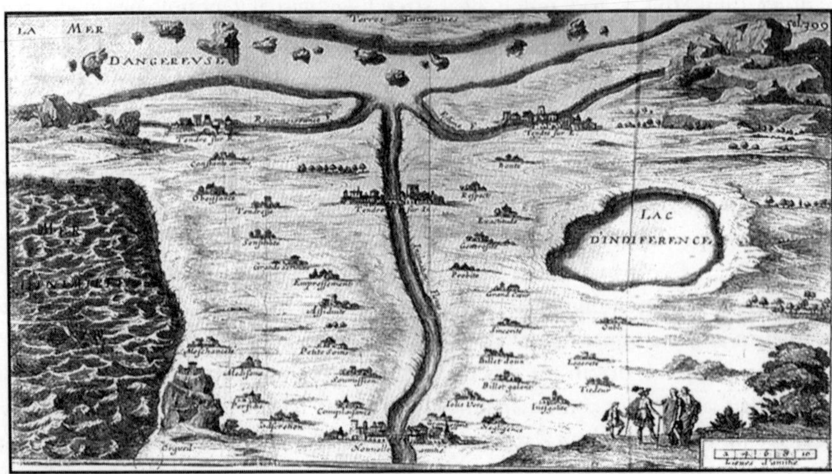

Figure 8.2 Madelaine de Scudéry's *Carte du pays de Tendre*, engraved by François Chauveau, 1654. Used with permission, Bibliotheque National de France

Tom Conley (2007) suggests that the map in *Clélie* might have been drawn in opposition to contemporaneous military cartographies, inaugurated by neo-Cartesian engineers under kings Henry IV through Louis XIV. These cartographers redrew the defensive lines of France and designed fortified cities in a time when new siege technologies were changing the ways of waging war. Conley goes on to point out that *Clélie* possibly reminded French society of the world of the *salon* and the space that women had crafted in opposition to the mechanistic world of warfare. In a similar vein, Bruno's arguments shift the context of visual geographies (her purview includes and goes well beyond maps) away from patriarchy and imperialism to a consideration of a tender geographical imagination.

With Bruno, we want to shift our argument to the intensification of emotional life that is possible through spatial images. At the very least we want to see a spinning of the Janus-headed coin of rationality and emotion so as to blur the boundaries between the sciences and arts that comprise geovisualization.

We begin to see that perhaps there is another side to the culture of digital immersive experience, and thus there is indeed a cultural politics that does not leave it to GIS and other mapping software to control every aspect of our digital engagement with geographical space. Affective geovisualization, we believe, provides geographers with the means to make more of GIS, to make something else of our digital engagements by reimagining how these engagements have the potential to work outside the terms that now dictate our levels of geovisual interaction and geovisual discourse. There are, however, certain aspects of the characterization of space that must be understood in order to realize the full potential of the geovisualization/affectivity engagement. First and foremost, it must be understood that all information has spatial qualities. This fact is critical to fully move from current geographical discourse (and the inability of GIS to embody 'information') and into an engagement that embraces the new visualizing technologies available to the discipline that allow the users and consumers of GIS to affectively realize spatial data. In addition, as Massey (1992; 2005) has advocated, space is dynamic and relational, and the relationships themselves create and define space and time; space is thus socially constituted, thereby setting the 'stage' for human affective engagement with digital spatial data. Massey further continues:

> Imagining space as always in process, as never a closed system, resonates with an increasingly vocal insistence within political discourses on the genuine openness of the future. It is an insistence founded in an attempt to escape the inexorability which so frequently characterizes the grand narratives related by modernity. (2005: 11)

We contend that current theories of geovisualization still adhere to a concept that fails to consider *how* users and consumers perceive, embody, experience, and thereby constitute information.

PLANNING DEEPLY: AFFECTING CHANGE THROUGH GEOVISUALIZATIONS

If you agree with the premise we made earlier that geovisualization technologies promise, at least to a degree, an altered digital reality through which spaces and places may be viewed anew, then it is not far fetched to follow our argument to its problematically utopian ends. Utopian visions of the world are not necessarily bad things (Harvey, 2000), but when they fail in practice they leave trails of betrayal, disappointment, frustration, and resentment. Paul Knox (1991: 187–8) points out that the US planning profession lost a good deal of its moral authority because of its inability to deliver utopian communities and, as a result, it became more aligned with the development and real estate industries with their penchant for rendering utopian images. This alliance led to a commodification of planning functions, which, in turn, facilitated the privatization and decentralized governance of urban life described by McKenzie (1994) and McCann (1995), and the bourgeois utopian desire to secure isolated and protected comforts through propertied individualism as described by Harvey (2000). Harvey points out that no matter how well intended, visualized utopias, like Thomas More's original conception, are tightly circumscribed spatially and morally. The internal spatial ordering of More's island strictly regulated a stabilized and unchanging set of social processes through, amongst other things, a series of exclusions (Harvey, 2000).

The spectacle of a utopian world anesthetizes us to the messiness of social relations. The spectacular waterfront of walkways and open space between high-rise hotels and condominiums promised to San Diego by the Manchester Financial Group, described earlier, belies the requisite exclusion of the poor and nearly poor from downtown areas and the creation of more land uses that curtail affordable housing. The artists' renderings presage a rational order of growth and development that imperiously garners public support while foreclosing upon the discussions, dissent, and consensus building from which public will is more likely to arise. The argument here is that although planning decisions are informed by mechanistic spatial models and empirical data represented through colorful cartographic and digital displays, more often than not an actual decision – indeed the very heart of planning practice – is made during face-to-face meetings where emotions run hot and cold, where there is consensus building, where compromises are made (Aitken, 2002; Aitken and Michel, 1995).

Unfortunately, dissent is often politicked away by the seeming rational portrayal of spectacular maps, aerial photographs, and visualized data that anesthetize sensibilities and foreclose upon debate by turning citizens into imperial subjects.

And yet, another narrative, a counter-topography if you will, arises through affective geovisualizations. As we are absorbed into geographic images, it is possible also to be affected by the power of those images to highlight social and spatial injustice. Affective geovisualizations are soulful; they tug at our hearts to the extent that we may be mobilized to action. Showley's (2006) article, for example, breaks through the placeless and timeless contexts of San Diego's waterfront development suggested by the Manchester Financial Group to suggest deeply geographical and historical contexts for the current debate. Three further examples help illustrate visual contexts from which deep planning may arise.

The first is a focus on representation as a tool that combines qualitative and quantitative data, and an empirical predilection that begins from the ground up to foreground the power of flexible and fluid bottom-up analyses and visualization. LaDona Knigge and Meghan Cope (2006) outline an integrated and analytic methodology for bottom-up visualizations that elaborate cartographic connections for real people in real places, who want to effect change and inform planning processes. Using a community gardens project in Buffalo, New York, as an example, they outline the compatibility between visualization techniques and grounded theory's ethnographic methods. Although they do not make affect explicit in their integrated method, their conjoining of feminism and human ecology with everyday community garden narratives suggests the importance of emotional ties to visualized experiences (Figure 8.3).

The importance of Knigge and Cope's multiple representations for affective geovisualization is the creation of embodied histories and geographies that serve a local place. Oral histories contextualize different people in distinct places over time; maps ground and enliven lived experiences. There is no prediction or explanation here: a 'commitment and sensitivity to multiple subjectivities, meanings, and discourses, and to the relationships between context, power, and knowledge' (2006: 2035) engender constant mutations.

The second and third examples are more overtly emotional, pulling from queer theory as an epistemological basis for elaborating sexuality and desire within a multidimensional, socially constructed fluid space. Michael Brown and Larry Knopp (2008) argue that, like most other poststructural theories, queer theory stresses that knowledge is always produced in contexts of power and that representation is always partial, mediated, and political. By focusing on an ethnographic cartography of queer spaces in Seattle, they push visualization techniques through specific contexts of power. The outcome of their action-research project is a map of historically significant sites in Seattle for the Northwest Lesbian and Gay History Museum Project. A key factor in their map making, beyond grappling with contexts of power and contingencies of facts and truths, was trying to represent the unrepresentable.

Similarly, Marie Cieri (2003) counterposes 'being' and 'looking' in her elaboration of queer social space in Philadelphia. The work is articulated as a political and personal journey through lesbian and social space, and as with our previous two examples, Cieri focuses on different ways to create, engage, and portray geographical information. Central to her effort is a desire to 'offer new and potent ways of telling geographic stories that emanate not so much from "authoritative" sources such as government officials, planners, marketers, the news media and the geographic mainstream as from

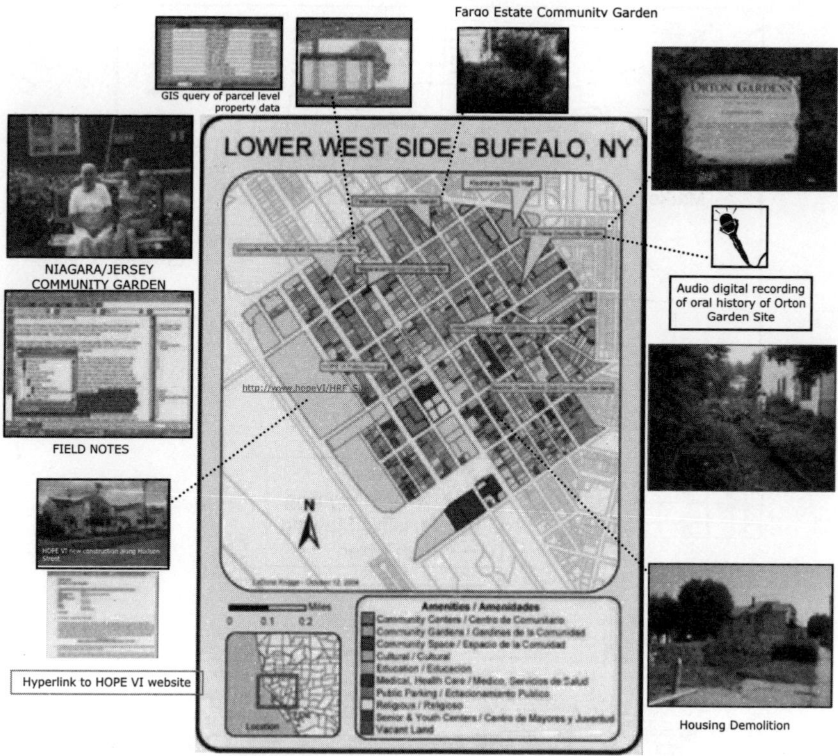

Figure 8.3 Everyday embodiments in Buffalo's Lower West Side. Used with permission, Knigge and Cope (2006)

populations themselves that are generally studied and represented by those authorities' (2003: 147). There is a fluidity to Cieri's maps (Figure 8.4) that reflects the complexity with which lesbian bodies occupy, transgress, and transform Philadelphian space. Although she calls this a 're-representation' it is also an affective geovisualization that pulls viewers away from standard tourist guides and into, and through, queer lives. Her maps embrace a desire to participate that is emotively drawn around issues of safety and curiosity as well as the desire to encounter people of similar identity.

In the same way that Scudéry's seventeenth-century tender mapping is in tension with royally appointed contemporaneous French map makers, so too Marie Cieri, Meghan Cope, LaDona Knigge, Michael Brown, and Larry Knopp create works that collide with hegemonic map-making epistemologies. And like Scudéry, none of the map makers is a traditionally trained cartographer or GIS specialist. Herein arises the hope of subversion and a reworking using accessible and affordable visualization technologies.

CONSTANT MUTATIONS

The three examples that close our discussion highlight grounded and queer forms of affective geovisualization that eschew technological advances in representation for

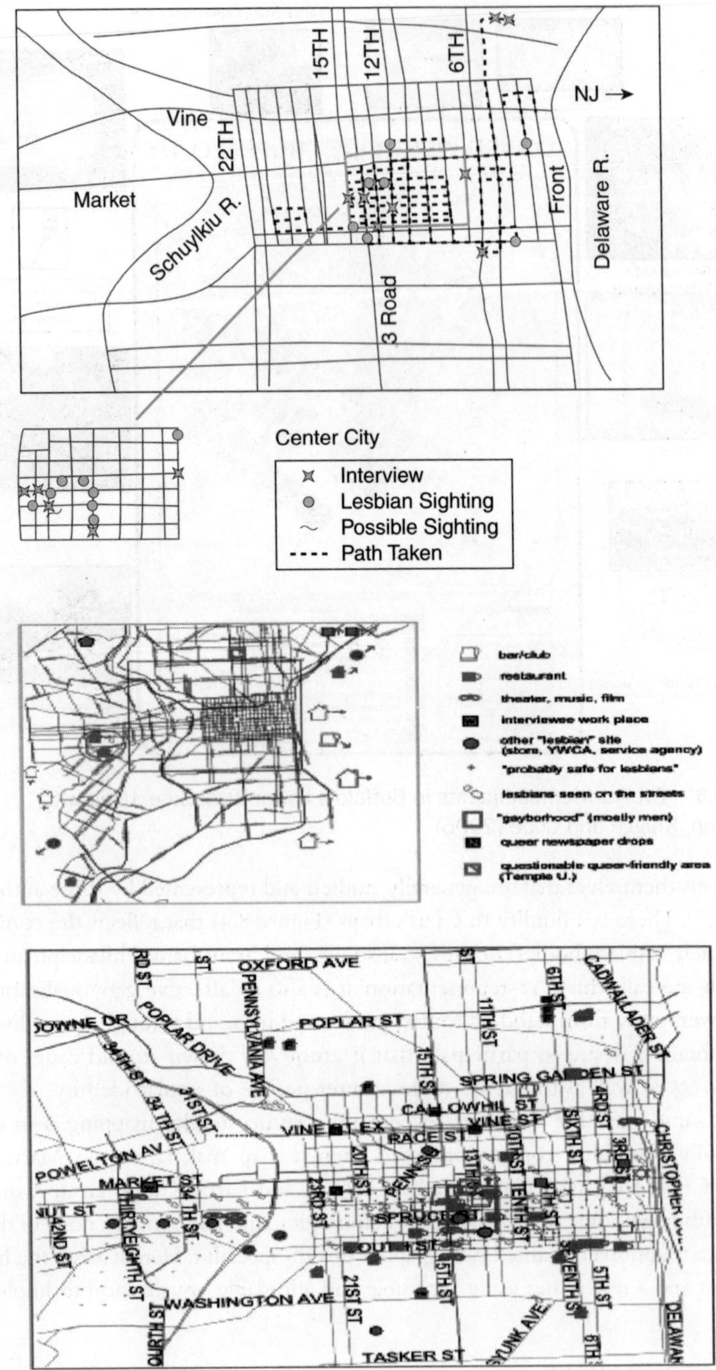

Figure 8.4 Lesbian perceptions of queer space in Philadelphia. Used with permission, Cieri (2003)

emotionally contextualized and politically charged visual statements that invite debate. There must be, as Harris and Harrower believe, a 'fuller engagement between debates in critical cartography and broader discussions from social theory' (2006: 2); and we believe that a deeper grasp of the affective properties of GIS offers geographers a greater range of opportunities to explore the possibilities of cartographic display within the context of new digital and multimedia technologies.

Geographers must explore beyond standard data presentation and dig deeper into the virtual structures and affective poststructures of GIS in its digital and virtual totalities, not only using standard geovisualization techniques and technologies but also drawing upon new social and critical theories. We accomplish this interaction in our examples: the San Diego planning controversy provides the modality for an affective geovisualization that is provided textually by the Showley critique. In addition, a more knowledgeable engagement of GIS completes the task of providing the user access to geospatial information and knowledge in a manner that is best suited for the user's cartographic and visualization skills. This takes geographical knowledge out of the private realm and into the public domain, as suggested by Knigge and Cope (2006), where it can be easily and quickly accessed through the use of a specific map metaphor that the user will enjoy utilizing. In addition, by understanding more fully how the digital virtual data are actualized socially and politically, geographers can more readily take advantage of new actualizations such as those created by Brown and Knopp (2008) and Cieri (2003).

It is our hope that affective geovisualization, as a new form of engagement, can bridge the gap between GIS-based cartographies and material geographies. We believe that affective geovisualization enables geographers to rethink the relationship between human users/consumers and digital information. Affective geovisualization opens up the potential for geographers to interface more effectively with digital datasets. Additionally, affective geovisualization opens the realm of the visual in ways that transcend the strict governing binaries of digital code. We believe that our form of critical analysis denaturalizes previous geography/GIS discourse, especially ostensibly objective technologies, for observation and classification in the same way that GIS and remote sensing also act as technologies for value and power. GIS is an excellent subject for geovisual studies because we can take into consideration their philosophical grounding, their history, and their production. Thus, we can then use affective geovisualization to suggest alternative orderings of knowledge. Geovisualization can take GIS representations that are presented as natural, universal, or true and analyze them so that alternative narratives, based on geography, become visible. We can also gain some understanding of how mainstream GIS users set themselves as examples to recognize and follow; we can then understand the political interests underlying the production of these geographic representations by using affective geovisualization to study their transparency. Geovisualization also promotes the look and being of the viewer/consumer – understanding that production comes first, followed by the perception it guides. We can thus be exposed to the interconnections between private and public, especially the private meanings and uses in memories and family histories and the visual tools already in place there; and, finally, we can also expose the interconnections between the physical and its digital GIS counterpart. By gaining an understanding of the affective nature of GIS, geographers can begin to understand what until now has been considered indiscernible. Geographical representations change meaning as the environment changes, so that the function of visual characteristics in relation to social processes with the GIS environment can be the purveyor of specific relations to bodies that are in continual flux. These characteristics install emotional comfort or distancing, confinement, intimacy or threat, but also, as a cognitive mode of understanding, they provide a 'scientific' method for grasping the complexities of our world. In addition, and most basic, the intertextual relationships between GIS, as a series of data-based images, and our different participating senses, the affectivity of the image in other words, require a new form of geovisual analysis. Finally, following Hooper-Greenhill (2000), affective geovisualization can work towards a social theory of visuality, focusing on questions of what is made visible, who sees what, and how seeing, knowing, and power are interrelated. Affective geovisualization can thus be used to examine the act of seeing as a product of the tensions between external images and internal thought and cognitive processes and passion.

REFERENCES

Acevedo, William and Masuoka, P. (1997) 'Time-series animation techniques for visualizing urban growth', *Computers & Geosciences*, 23 (4): 423-35.

Aitken, Stuart C. (2002) 'Public participation, technological discourses and the scale of GIS', in William Craig, Trevor Harris and Daniel Weiner (eds), *Community Participation and Geographic Information Systems*. London: Taylor and Francis. pp. 357–66.

Aitken, Stuart C. and Craine, Jim (2006) 'Visual methodologies: what you see is not always what you get', in Robin Flowerdew and David Martin (eds), *Methods in Human Geography*. Harlow: Prentice-Hall. pp. 250–69.

Aitken, Stuart C. and Michel, Suzanne (1995) 'Who contrives the "real" in GIS? Geographic information, planning and critical theory', *Cartography and Geographic Information Systems*, 22 (1): 17–29.

Anderson, Kay and Smith, Susan (2001) 'Emotional geographies', *Transactions of the Institute of British Geographers*, 26: 7–10.

Barthes, R. (1971) *Mythologies*. London: Cape.

Benjamin, Jessica (1986) 'A desire of one's own: psychoanalytic feminism and intersubjective space', in Teresa do Lauretis (ed.), *Feminist Studies/Critical Studies*. Bloomington, IN: Indiana University Press.

Brown, Michael and Knopp, Larry (2008) 'Queering the map: the productive tensions of colliding epistemologies', *Annals of the Association of American Geographers*, 98 (1): 40–58.

Bruno, Giuliana (2002) *The Atlas of Emotion: Journeys in Art, Architecture, and Film*. New York: Verso.

Buttenfield, Barbara P., Gahegan, M., Miller, H. and Yuan, M. (2000) 'Geospatial data mining and knowledge discovery', http://www.ucgis.org/priorities/research/research_white/2000%20Papers/emerging/gkd.pdf, 22 April 2007.

Cieri, Marie (2003) 'Between being and looking: queer tourism promotion and lesbian social space in Greater Philadephia', *ACME: An International E-Journal for Critical Geographies*, 2 (2): 147–66.

Conley, Tom (2007) *Cartographic Cinema*. Minneapolis: University of Minnesota Press.

Connolly, William (2002) *Neuropolitics: Thinking, Culture, Speed*. Minneapolis: University of Minnesota Press.

Crang, Michael, Crang, Philip and May, Jon (eds) (1999) *Virtual Geographies: Bodies, Space and Relations*. New York: Routledge.

Cresswell, Tim and Dixon, Deborah (eds) (2002) *Engaging Film: Geographies of Mobility and Identity*. New York: Rowman and Littlefield.

Damasio, Antonio (1994) *Descartes' Error: Emotion, Reason, and the Human Brain*. London: Penguin.

Davidson, Joy, Bondi, Liz and Smith, Mick (2007) *Emotional Geographies*. Aldershot, UK: Ashgate.

Debord, Guy (1983/2000) *The Society of the Spectacle*. Detroit: Black and Red.

Dykes, Jason A. (1997) 'Exploring spatial data representation with dynamic graphics', *Computers & Geosciences*, 23 (4): 345–70.

Fabrikant, Sarah (2005) 'Towards an understanding of geovisualization with dynamic displays: issues and prospects', in T. Barkowski, C. Freska, M. Hegarty and R.K. Lowe (eds), *Proceedings of Conference on Reasoning with Mental and External Diagrams: Computational Modeling and Spatial Assistance*, Stanford, CA. pp. 6–11.

Fyfe, Nick (ed.) (1998) *Images of the Street*. New York and London: Routledge.

Gahegan, Mark and Pike, William (2006) 'A situated knowledge representation of geographical information', *Transactions in GIS*, 10 (5): 727–49.

Haraway, Donna (1989) *Primate Visions: Gender, Race, and Nature in the World of Modern Science*. London: Routledge.

Haraway, Donna (1991) *Simians, Cyborgs and Women: The Reinvention of Nature*. New York: Routledge.

Harris, L. and Harrower, M. (2006) 'Critical interventions and lingering concerns: critical cartography/GISci, social theory, and alternative possible futures', *ACME*, 4 (1): 1–10.

Harvey, David (2000) *Spaces of Hope*. Berkeley, CA: University of California Press.

Hooper-Greenhill, E. (2000) *Museums and the Interpretation of Visual Culture*. London: Routledge.

Knigge, LaDona and Cope, Meghan (2006) 'Grounded visualization: integrating the analysis of quantitative and qualitative data through grounded theory and visualization', *Environment and Planning A*, 38: 2021–37.

Knox, Paul (1991) 'The restless urban landscape', *Annals of the Association of American Geographers*, 81 (2): 181–209.

MacEachren, Alan M. (in collaboration with Buttenfield, Barbara P., Campbell, James B., DiBiase, David W. and Monmonier, Mark) (1992) 'Visualization', in *Geography's Inner Worlds: Pervasive Themes in Contemporary American Geography*. New Jersey: Rutgers University Press. pp. 101–37.

MacEachren, Alan M. (1994) 'Visualization in modern cartography: setting the agenda', in A.M. MacEachren and D.R.F. Taylor (eds), *Visualization in Modern Cartography*. Oxford, UK: Pergamon.

MacEachren, Alan M., Gahegan, Mark and Pike, William (2004) 'Visualization for constructing and sharing geo-scientific concepts', *Proceedings of the National Academy of Sciences*, 101 (1): 5279–86.

Massey, D. (1992) 'Politics and space/time', *New Left Review*, 196: 65–84.

Massey, D. (2005) *For Space*. London: Sage.

McCann, Eugene (1995) 'Neotraditional developments: the anatomy of a new urban form', *Urban Geography*, 16 (3): 210–33.

McKenzie, Evan (1994) *Privatopia*. New Haven, CT: Yale University Press.

Ramachandran, V.S. (2008) *Mirrors in the Mind: The Science of What Makes Us Human and Creative*. In press.

Robins, Kevin (1996) *Into the Image: Culture and Politics in the Field of Vision*. London: Routledge.

Rogin, Michael (1990) '"Make my day!": spectacle as amnesia in imperial politics', *Representations*, 29: 106.

Rose, Gillian (1993) *Feminism and Geography*. Oxford: Polity.

Sauer, Carl (1956) 'The education of a geographer', *Annals of the Association of American Geographers*, 46: 287–99.

Showley, Roger (2006) 'On the waterfront: battle lines drawn', *The San Diego Union–Tribune*. May. E–2, archived at http://www.signonsandiego.com/uniontrib/20060507/news_1h07 smokenew.html.

Skupin, André and Fabrikant, Sarah Irina (2003) 'Spatialization methods: a cartographic research agenda for non-geographic information visualization', *Cartography and Geographic Information Science*, 30 (2): 95–115.

Thrift, Nigel (2004) 'Intensities of feeling: towards a spatial politics of affect', *Geografiska Annaler B*, 86: 57–78.

Wasko, J. (1995) *Hollywood in the Information Age*. Austin, TX: University of Texas Press.

Wylie, John (2006) 'Poststructural theories, critical methods and experimentation', in Stuart Aitken and Gill Valentine (eds), *Approaches to Human Geography: Philosophies, People and Practice*. Thousand Oaks, CA: Sage. pp. 298–310.

9

TOWARDS A GENEALOGY OF QUALITATIVE GIS

Matthew W. Wilson

ABSTRACT

By situating qualitative GIS research among the various research trajectories in the 'GIS and society' tradition, I argue that qualitative GIS enacts a specific researcher positionality. In reading this positionality against the original call for a 'critical GIS', I argue that qualitative GIS must continue to problematize its relationship with bits of code and practices of coding. This is a call for a situated qualitative GIS, a genealogical tracing of the multiple rootings of this research and development endeavor. Situating qualitative GIS as a critical GIS, I consider how our relationships with technologies create concerns around the positionality of critique. This is a question of 'insiderness', answered through the call for a kind of insider gaze, which is privy to and constitutive of the terms and terminologies of the technology. To further elaborate how the positionality of qualitative GIS research differs from other work in GIS and society, I discuss three subfields: STS studies, ethno(carto)graphies, and socio-behavioral studies of GIS. Qualitative GIS is distinct in its perspective, what I argue is a techno-positionality. I further discuss how this techno-positionality enacts knowledge production differently, to begin to situate qualitative GIS research, to allow it to speak back to the earlier disciplinary debates about GIS, and to begin to think genealogically about qualitative GIS.

INTRODUCTION

> While human geographers are familiar with the historically contingent nature of positions within debate, there is little practical awareness of fundamental limitations that constrain the 'paying of attention' from within a particular position. (Hannah and Strohmayer, 2001: 400)

Like any scholarly contribution to academic literatures, technological innovation can be situated within previous inventions and interventions. Innovation requires our imaginations of futures nearly present and readings of pasts not entirely departed. Qualitative GIS is one such innovation, and this volume speaks to the range of thought currently working the boundaries of what is considered geographic information systems. Qualitative GIS is, in this sense, both a speaking back to the technological and literary traditions in GIS and a marked departure. This chapter argues that qualitative GIS emerges out of debate, out of the suspicions of disciplinary irrelevance, and out of accusations of reductive empiricism. Our abilities to proceed in qualitative GIS endeavors must be read through these debates, through the anxieties surrounding GIS in the 1990s. As Hannah and Strohmayer discuss in the epigraph, debates in human geography are often aggregated into the positions of the debaters (the regional school versus the quantitative school), which elides the various nuances and differences that articulate the debate. Instead, we resort to epithets – 'critical geographers', the 'GISers', etc. – glossing the possibilities for commonality, and disenabling those left at the borders. This chapter anticipates a genealogy of qualitative GIS, to draw in the various discourses that contribute to its emergence, and continue to challenge its pursuit. The point is to be attentive to how our present relationship with geographic information technologies has been enabled through specific inventions and interventions. More specifically, to understand qualitative GIS, we must consider the lines of debate in critical GIS that have led us here.

While preparing this chapter, I was fortunate to attend Enrique Chagoya's *Borderlandia* at the Des Moines Art Center in Des Moines, Iowa. His irreverence to modernist art form and his dramatic use of incongruent elements, typified in works like *Le Cannibale Moderniste* (1999), create a dialogic space for contestation. Against a mock, tranquil Monet background, Chagoya sketches a bleeding and armless Picasso, running from an African woman, baby in tow, feeding on Picasso's arm, while being watched by the floating, severed head of Monet, who is quoting a Mondrian piece. A Disney character sprints away from the scene. Chagoya's technique presents for me a kind of visual metaphor for the spaces constituted by qualitative GIS: spaces of irreducibilities and contestations. Similarly through surprising juxtapositions and artful incongruities, the qualitative GIS of Jung (this volume) and Kwan (2007) enable these kinds of spaces – to challenge traditional understandings of GIS and model new representative practices.

In relating *Borderlandia* and qualitative GIS, I intend to question the work that each performs. What kinds of subjects are assumed and produced through their gaze? What roles are enacted by the viewer and the artist-technician? While Chagoya vividly displays his politics, there are more subtle, political interventions at work. To recognize these disruptions, a viewer of his work must recognize the constitutive elements (the disfigured, modernist painter and the revengeful objects of modernist paintings) to realize their incongruence. It is in this sense, of inquiry about the viewer's positionality, that I want to push back on qualitative GIS. To realize the implications of this subdisciplinary field, the viewer must recognize how what went before is being radically altered. These are the questions at the borderlands: to inquire about relations of insiderness and outsiderness in critical praxis, to ask what is enabled by this tension. In order to conceptualize the implications for qualitative GIS, I argue that a genealogy of qualitative GIS must be linked up to critical GIS: an interrogation of the questions and juxtapositions this movement attempts to address and to allow. Doing so not only contributes to our understanding of what is 'mixed' in mixed methods research, but also further considers the implications of this mixing. I choose to begin my telling of qualitative GIS here, as a question of how and who gazes inside the (Arc)toolbox.

EMERGING QUESTIONS OF 'INSIDERNESS'

Questions of insider and outsider relations *vis-à-vis* GIS have enabled a series of productive critiques in the form of critical GIS, participatory GIS, and qualitative GIS. This questioning has led some in critical GIS to link critique to internal positions that are technically informed, most specifically in the work of Nadine Schuurman. Schuurman (1999) constituted 'critical GIS' as a critical engagement *with* the technology, on its own terms and in its own terminologies. This was a research positionality concerned with 'insiderness'. The way in which qualitative GIS emerges as a viable field of study can be traced to the enabling responses Schuurman offers towards the 1990s critiques of GIS. In this sense, I am calling for a situated qualitative GIS, which is responsible to the critiques and contributions of work in critical GIS. In this section, I consider how critical GIS, as offered by Schuurman (1999) in her research monograph published in *Cartographica*, situates qualitative GIS. I offer two points of departure in this situating, arguing that a qualitative GIS situated as a qualitativeness of GIS theory and practice has implications both for critique and for critical praxis.

First, qualitative GIS can be considered a political intervention. In addition to being a host of technical and methodological considerations for 'out-of-the-box' thinking in GIS development and use, qualitative GIS is constructed by a series of academic debates around the fit of GIS within disciplinary geography. Those advocating a more qualitative GIS are potentially responding to an assumption that GIS is quantitative, or lacks a qualitative capacity – a dichotomy problematized by Pavlovskaya (this volume). Pavlovskaya does not insist that GIS is either quantitative or qualitative; rather, she argues that GIS was always non-quantitative. Her claims to an originally non-quantitative GIS enable, for her, a powerful rethinking of the practice and theorization of GIS. This kind of discussion of the quantitativeness and qualitativeness of GIS is indicative of the kinds of tensions which are important in Schuurman's (2000) critical GIS – to be concerned with the technology on its own terms, in its terminologies. Qualitative GIS, like critical GIS, also serves to remind critical geographers of the importance of *doing* GIS in critical contexts, extending calls for a critical cartographic literacy (Harris and Harrower, 2005; Johnson et al., 2005). Qualitative GIS inherits this tradition, via critical GIS, and enables this critical praxis. By reading qualitative GIS through the lens of critical GIS as it emerges in debate and from contestation, I introduce the second point of departure in situating qualitative GIS – the construction of an *insider gaze* that is privy to and constitutive of the terms and terminologies of the technology.

Second, an insider gaze is central to critical GIS, and enables, I argue, the more recent move to qualitative GIS. I need to preface this claim, however, with the recognition that the critical GIS first articulated by Schuurman in 1999 has now mutated into various forms taking up different dimensions of the 'critical' (qualitative GIS among them). This earlier critical GIS, as bounded by Schuurman's research monograph, is where this insider gaze is manifested. It emerges as a counter-argument to a certain form of critique, whereby the critic is challenged for their positionality in relation to the technology. Schuurman, in her history of GIS critique, provokes a rejoinder to earlier critics of GIS. In her remarks concluding a review of the GIS debates, she worries that critique which employs language outside the terminology of GIS will not 'gain the ear of GIS researchers' (2000: 587). Critiques of technology must remain 'relevant to the technology', she writes, and to do so, critics of technology must acquire the 'vocabulary of the technology' (2000: 587). Schuurman and Leszczynski (2006) take this call up in later research about metadata standards, highlighting the importance of ontology-based metadata. Here, they investigate the inner workings of these systems, to remain responsible to the issues of the technology.

However, the operative field of critical GIS is constituted through Schuurman's language. She recognizes other forms of critique as in danger of becoming or remaining 'peripheral', and further insists that social theory can 'only tangentially engage a technology which is written in the language of computational algebra and constructed through the laws of physics' (2000: 587). Through the use of territorial markers like 'periphery' and 'tangents', she encourages a policing of the border separating relevant and internal critique from peripheral, tangential, and exterior critique. Her insistence on internal critique is enabled, she argues, through this 'constitutive outside'; in other words, her strategy is precisely to code certain critics and their critique as unsatisfactorily formed with regard to the technology (1999: 10). This typing of critical positions is further reinforced in her collaborative piece with Pratt. Here, Schuurman and Pratt (2002) write in the ethic of Pratt's (1996) earlier piece against 'trashing' as a style of critique. Again, their issue is with the degree of internalness of the critic – a degree measuring the critic's 'care for the subject', in this case, 'the future of the technology' (2002: 295). Their call for a new attitude in GIS critique, as distanced from the 'morally and intellectually superior outsider', constitutes a curious position, defined only in opposition to some differently coded exteriority. Table 9.1 demarcates the boundary between inside and outside, extends these arguments, and lexically casts the strategic field for internal critique in early critical GIS. Internal critiques rely upon an insider gaze and enforce a membership that is more closely aligned with/for the technology, and requires certain technical proficiencies (or at least terminologies). From this vantage point, the critique becomes legitimate and relevant, or simply more palatable.

Feminist geographers are attentive to discussions of internal critique. Kim England (1994; 2002) discusses how our own understandings of self mitigate our abilities (or inabilities) to conduct research. Lynn Staeheli and Patricia Martin (2000) point to the blurring of boundaries between the field and the researcher, and the power relations that underlie this relationship. Staeheli and Martin write, 'positioning oneself in relation to the field … become[s] [an] immanently political [process]' (2000: 145). This positioning has been described as 'in-betweenness' (Katz, 1994; as cited by Staeheli and Martin, 2000: 146) and as 'borderlands' (Marshall, 2002). Marshall considers insider–outsider tensions a borderland in her ethnographic research; her positionality as the researcher was multiply defined with various human relationships in the study. The status of being *inside to* or *outside of* defined the moments of research, further nuancing Marshall's narrations. In this way, Marshall places insider–outsider relations in dialog in her ethnography. Likewise, Gillian Rose (1997) proposes a reflexivity that problematizes the distance constituted by various borderlands, including insider–outsider positions (and their permutations: see Table 9.1).

Table 9.1 Insider–outsider discourse in early critical GIS

Insider	Outsider
Vocabularies of the technology	Vocabularies of social theory
Care for the subject	Care for the critic
Bolstering	Trashing
Relevant	Irrelevant
Legitimate	Illegitimate
Constructive	Destructive
Positive	Negative
Proximate	Distant

These two points of departure highlight two ways we might situate qualitative GIS, as a troubling of the kind of political intervention undertaken and of the positionality of critique. I have discussed these departures as emerging from the work by Schuurman to define a critical GIS. More recent work in critical GIS/cartography contributes to new understandings of these departures, particularly new understandings of the 'critical'. The practicing of qualitative GIS might be understood as engaging a kind of double critique, echoing Crampton and Krygier (2005) in their discussion of implicit and explicit cartographic critique. Critique, they suggest, can be understood as both an interrogation of knowledge-making practices and an alteration of these practices in a way that affects change. Qualitative GIS is emblematic of these new understandings, and yet, as I shall discuss, the way in which the 'insider gaze' constitutes its field of operation is cause for further genealogical investigation. We must ask, therefore, what kinds of positionalities are afforded this gaze, and what are the implications for this kind of critical, qualitative GIS? To get on with this mode of questioning, I situate the positionalities of 'GIS and society' research more broadly in the following section.

POSITIONALITIES IN GIS AND SOCIETY RESEARCH

By maintaining that certain critical positions wage more relevant GIS critique, Schuurman (1999) constitutes a site to place her research – what she termed critical GIS – and further invokes a way of thinking about technologies and critique (see also Schuurman, 2006). As a subdisciplinary label, critical GIS marks research which seeks to critique technology through reconstruction – by engaging the technology on its own terms. Schuurman and Pratt (2002) point to examples of research by Sarah Elwood, Mei-Po Kwan, and Sarah McLafferty, where GIS is framed as a non-neutral tool in state–community dynamics and where GIS is mobilized as a tool to interrogate these more material dynamics. These research projects, they argue, are emblematic of a kind of constructive critique that recognizes the transformative potential of the technology itself – a potential recognized through the insiderness of this form of GIS critique. This, as I discussed previously, is a kind of positionality, and one I shall return to in the next section. There are other positionalities in GIS and society research, which take up different relations with the technology. In order to situate qualitative GIS, I discuss how related branches of GIS and society research enact certain research positionalities in relation to the technology. These positionalities crystallize around specific technology–researcher relationships, or the degree to which the researcher actively reworks the technology as the means or the end point of the research. Here, I propose four clusters of GIS and society research and discuss their specific positionalities: science, technology, and society (STS) studies of GIS; ethno(carto)graphic studies; socio-behavioral studies; and qualitative GIS.

STS studies within GIS research use discourse analysis and actor-network theory to problematize the power–knowledge relationships between science, technology, and society. This research draws upon Latour (2005) and Haraway (1991) to articulate the interconnectivities of technology and society, to historicize their co-embeddedness (e.g. Chrisman, 2005; Curry, 1998; Ghose, 2007; Harvey and Chrisman, 1998; Pickles, 1997; 2004; Poore and Chrisman, 2006). This approach, emerging out of the GIS critiques of the mid 1990s, draws upon the perspective of science and technology to historicize GIS. For example, Poore and Chrisman (2006) advocate a retheorization of geographic information. Their method is one of cautious historicization of metaphors about information: the first being a 'metaphor of invariance', and the second a 'metaphor of refinement'. The former metaphor, they argue, informs the cartography-as-communication tradition, while the latter describes an analytical approach to the production of refined information and eventually wisdom. Here, Poore and Chrisman draw upon STS studies to reveal how these metaphors enabled certain cartographic and GIS practices. They conclude with a social theory of geographic information, acknowledging that 'information is actively transformed and reworked by its recipients' (2006: 520). Their approach, similar to Latour's actor-network theory, is to place these metaphors in relations with other knowledge productions, theorizations, and practices and to challenge the dominant metaphors by revealing what they necessarily hide and disallow. Their position as researchers is to intervene through renewed storytellings of geographic information systems, of their origins and implications.

This form of critique is not an admonishment of GIS; rather, this work seeks to place GIS in broader narratives of global capital, institutional networks, and information sciences. When employing discursive and actor-network analysis, the researcher negotiates a position that actively rereads the histories and implications of technology. This positionality is about reading the technology through various (social) theoretical perspectives, be they poststructuralist or historical-materialist. Schuurman (2000) seems most troubled by this positionality in research. Similarly Leszczynski problematizes this research positionality as one concerned with discourse and not, she argues, with the more relevant 'empirical level of the machine' (2007: 74). STS studies of GIS, from the perspective of the insider advocate, may be criticized as being too distant from the code-based realities of the technology. Their positionality, in other words, remains too divested from the mechanisms of the technology.

It is helpful to read STS studies in juxtaposition with ethnographic studies. Ethnography, or within GIS and society what I term ethno(carto)graphy, is a method popularized due to the belief that ethnographies appropriately place the researcher *within* the process of doing research. Here, researchers attempt to correct or ground their inquiry *in* the phenomena of their study (Herbert, 2000; Marshall, 2002; Matthews et al., 2005; see also Knigge and Cope, this volume). It is also a methodology of collaboration, or of participatory research (Benson and Nagar, 2006; Brown and Knopp, 2008;

Elwood, 2006; Pain, 2004; Williams and Dunn, 2003). As an example of ethno(carto)graphy, Sarah Elwood researches the tactics of community organizations with which she collaborates, analyzing the discursive formulations of 'spatial narratives' or 'the production of spatial knowledge' (2006: 332). She argues that findings in the GIS and society literature are disparate due to dialectical battles between various interpretations of community use of GIS, as 'cooptation or resistance, an activist role or a service delivery role, expert knowledge versus experiential knowledge' (2006: 326). Her methodology is a qualitative analysis of the discourse produced by these agencies' maps and discussed in interviews she conducts. These discursive formulations include various narratives which, Elwood argues, are necessarily spatial: narratives of needs, assets, injustice, accomplishment, and reinterpretation (2006: 332).

Ethno(carto)graphic methodologies are both ethnographic and cartographic – a production of critique through the discursive work that collaborative cartographies enable. The positionality of the researcher in ethno(carto)graphy is explicitly invoked in the process of doing research. Elwood's positionality as a 'technical expert' in relation to the organization bolsters her analysis and findings. Her ability to perform research, therefore, is mediated through this positionality – of being marked both as an insider *and* as an outsider to the process of spatial knowledge production. The ethnographer's implicated position of being a participating observer is one requiring cautious introspect and reflection (Marshall, 2002). Ethnography in GIS and society, as Elwood (2006) demonstrates and Elwood and Martin (2000) discuss, requires an exposed consideration of place and incorporates critical map and critical GIS readings. Ethno(carto)graphy necessitates this positionality of preoccupation with interiority and exteriority as productive of critical research practices.

By some accounts, socio-behavioral studies of GIS maintain a close relationship to the technology. This relationship is one that constructs and configures the codes and practices of the technology itself (Dragicevic and Balram, 2004; Jankowski and Nyerges, 2001; Nyerges et al., 2006; Peng, 2001). In contrast to STS studies of GIS and ethno(carto)graphies, these studies base much of their legitimacy and relevancy on this relationship. Tim Nyerges et al. (2006) deploy a conceptual framework for analyzing decision-making situations which use software to support decisions about the management of water resources. They argue that ethnography, participant observation, and case studies cannot fully interrogate the use of GIS. Instead, they propose data collection about human–computer–human interaction (HCHI), in what they describe as a socio-behavioral study 'at fine levels of resolution in participatory processes' (2006: 705). To code these processes, the team videotaped the decision-making situation as the software they developed was used. They then coded the sequences on the videotape as it matched the research team's scheme of a 'macro–micro decision process', or general and specific categories of a decision process. Their conceptual framework provided a way to organize their research findings, while their positionality as software developer and 'technical expert' further enabled their proposals for new technical and procedural fixes to a decision-making problem.

From the perspective of the insider gaze, the socio-behavioral researcher positionality is intensely insider. The command and control of the technology in research situations allow these researchers to build technological agendas for replication and extension. The socio-behavioral researcher makes use of scientific distance in this way, underlining the separation between researcher and subject – all while requiring the insider positionality of technical proficiency. This research is often not framed as critique, due to its own ambivalences about relevancy to the terminologies of the technology. And yet, socio-behavioral research engages the technology in a way that enacts the kind of technological engagement that Schuurman advocates. This presents a problematic for early critical GIS as presented by Schuurman (1999), as socio-behavioral studies of GIS are usually missing from the critical GIS typology of GIS and society research.

Qualitative GIS, I argue, engages a different researcher positionality from the three categories of GIS and society research that I have previously discussed. Returning to the first section of this chapter, where I propose two departures for situating qualitative GIS, I enter qualitative GIS as distinct from other GIS and society research. And yet, it explicitly and implicitly inherits from these three traditions. From an STS perspective, qualitative GIS is a technology *being* situated within its institutional, capitalistic, and disciplinary histories. This perspective enables its recognition as a political intervention. From an ethno(carto)graphic perspective, qualitative GIS is a technology which recognizes the qualitativeness of collaborative knowledge production. It is a participatory action technology. From a socio-behavioral GIS perspective, qualitative GIS implicitly inherits the motivation to alter the technology and the techniques, to change the configurations and propose new specifications. It is a reconstructed technology.

These congeries of partial histories co-shape qualitative GIS. As this edited collection demonstrates, qualitative GIS envisions a qualitativeness of both the content of the GIS and the processes that shape it. In their introduction to a special issue on qualitative GIS, Kwan and Knigge (2006) highlight the border-smashing potential of qualitative uses of GIS, detethering quantitative from positivist epistemologies and qualitative from critical epistemologies (along these lines, see Lawson, 1995; Sheppard, 2001). Active in this account is the notion that research *with* and *on* GIS can invoke qualitative techniques of interpretation, visuality, reflexivity, and contextualization as well as more idiographic pursuits to situate knowledge and knowledge practices. Articles in that special issue examine the potential for GIS to be used critically, as an extension of qualitative research methods (Dennis, 2006; Knigge and Cope, 2006; Pain et al., 2006; also see Bell and Reed, 2004; Kwan, 2002), and further challenges the quantitative origins of GIS (Pavlovskaya, 2006; also this volume). Qualitative GIS, I argue, enables a kind of positionality that is attentive to the ins and outs of geographic information systems, and is motivated to complicate the rules and responsibilities of the software and its practice. Therefore, it is energized by the insider–outsider tension, and yet moves beyond this framing. It necessitates a techno-positionality of a conflicted insider, confronting colliding epistemologies and embracing incongruities. This *techno-positionality* is further taken up in the following section, as I reconceptualize the work of qualitative GIS and highlight the radical openness that those engaged in qualitative GIS seek to embolden.

TECHNO-POSITIONALITY IN QUALITATIVE GIS

By examining the positionalities produced in the different ways of doing GIS and society research, I have situated qualitative GIS in distinction. Qualitative GIS exceeds the insider–outsider tensions that constituted 'critical GIS', and additionally, enacts the researcher–technology relationship differently from other GIS and society research. In this section, I consider what is entailed by foregrounding the qualitativeness of GIS theory and practice, in the work of Jung, Kwan, and Knigge and Cope. Here, I draw on a notion of qualitativeness that is about the embeddedness of practice and the context of counting (Moss, 1995). It is a qualitativeness of fixity. As a system of representation, qualitative GIS necessitates a moment of fixity captured by the image and the database. One might assume that this fixity gives GIS its strength as a tool of generalization and exploration, and yet the question of what becomes fixed, when, and by whom, is a site of contestation. In this tone, qualitative GIS has the potential to work more as a system of re-presentation, by opening up these questions of what, when, and whom (as well as how, and of course where) to multiple authorings and re-creation. Fixity is nuanced here as temporary fixes. While qualitative GIS requires these moments of fixity, it enables a qualitativeness in doing so.[1] It offers an active recharacterization of the underlying logics in GIS. This is a techno-positionality, a research positionality neither of the historical-materialist, nor of the technical expert or embedded ethno(carto)grapher.

Techno-positionality is a positionality in conducting research that is simultaneously about and with the technology. It is 'techno' in the sense that its relationship with technology is hybrid – a taking up of the discourses and the technicalities of the machine. It is a way of *doing* research through 'machinic vision' (Johnston, 1999).[2] Furthermore, it is a way of *doing* technology as a craft – of practicing technocritique. Qualitative GIS invokes this techno-positionality to recognize how these technologies enable shifts in discourse, while actively reworking the technology to enable an openness to incongruity and irreverence that is productive of new forms of knowledge. As such, this techno-positionality is a conflicted insider – privy to the terminologies of the technology, and yet uninterested in the continuities of the technology. It is a way of relating to technology that is neither entirely inside nor outside, relevant nor irrelevant, constructive nor destructive, in the sense of an earlier, strategic critical GIS, as depicted by Table 9.1.

This 'conflicted insider' techno-positionality is steeped in the technicalities of GIS, and yet seeks resistive practices, new collusions, and irreconcilables to challenge GIS at the level of code. From this perspective, the technology (of hardware and software) is conceptualized as a site of opening, of the possibility for new encodings, interactions, and interpretations. Jin-Kyu Jung (2007; also this volume) and LaDona Knigge and Meghan Cope (2006; also this volume) each move 'inside' the technology to rework what is meant by surface (in Jung's fantastic mosaic of embodied landscapes) and by metadata (in Knigge and Cope's reversal of coordinate-laden ground and ground-laden coordinates). Each challenges what we imagine to be map-like, and yet also brings analytical tools and procedure to bear on these new visualizations. In effect, they have constituted new visualizings. Their qualitative GIS is about multiplicities and contingencies – about joining together previously separated objects and practices.

Knigge and Cope, and Jung, artfully connote a kind of *mashup*. Mashups popularly refer to the cojoining of two or more instances of a particular medium, or of media. For instance, the use of internet map applications for other purposes, beyond the original intent of the map application, constitutes a mashup (Miller, 2006). Mashups become a political intervention in qualitative GIS, as they refigure data, source, metadata, image, and anecdote. Knigge and Cope allude to this conceptualization by describing how, in combining quantitative data with ethnographic data and 'iterative, reflexive rounds of analysis', this refiguration happens and is thereby 'attuned to multiple subjectivities, truths, and meanings' (2006: 2035). While 'mashup' has not been used to describe qualitative GIS research, it is appropriate because the term draws in other efforts in geographic information science (both academic and non-academic), including research in volunteered geographic information (Goodchild, 2007), affective GIS (Aitken and Craine, 2006; and this volume; Kwan, 2007), and web development using Google Maps and Google Earth.[3] I site these efforts together as mashups to consider their potential relatedness. Volunteered geographic information (VGI) demarcates a new area of study, emerging from a specialist meeting held in December 2007 in Santa Barbara, California (see special issue of *GeoJournal, 72:* 3–4). Michael Goodchild (2007) refers to VGI as depicting the 'flood of new web services and other digital sources [that] have emerged [and] can potentially provide rich, abundant, and timely flows of geographic and geo-referenced information'. The concern here is with shifting datascapes, or new ways of creating, storing, manipulating, and analyzing geographic information. Mashups, a kin of VGI, elude our traditional ways of knowing and seeing. Similarly, Aitken and Craine (2006) and Kwan (2007) have explored alternative (non)representative practices with GIS, to detether GIS from its fixed usages, and further, as Kwan writes, to demonstrate her restlessness with these technologies' involvement in war, conflicts, and surveillance.

These new visualizings mash together different ways of knowing. Much has been written about the epistemology of GIS (cf. Brown and Knopp, 2008; Lake, 1993; Pickles, 1997; Schuurman, 2000; Smith, 1992), and while we can be certain that this is contested terrain, one reason for the appeal and excitement of qualitative GIS is that what qualitative GIS seeks to know challenges earlier epistemological critiques of GIS. Jung's (2007) collage of georeferenced images with systematized relations to a qualitative analysis engine and Kwan's (2007) artful renderings of triangulated irregular networks surprise us. This surprise, a grotesque discourse, enables an active rethinking of *how we know* with GIS. Traditional GIS is often assumed to be positivistic, enabling a separation of subject from object. This separation becomes untenable in these instances of qualitative GIS. The surprises that this sort of technology engenders are due to the shifts in knowing necessitated by such new creations, a kind of playful mimesis. This element of surprise emerges in part due to the breakdown of insider–outsider relations with GIS technologies; qualitative GIS reworks knowledge by using the tools in ways that exceed their original purposes. These multiple claims to knowing, while seemingly incongruous, are foregrounded in qualitative GIS. These incongruities surprise and reveal. Qualitative GIS is an exploration of these incongruities.

Qualitative GIS also challenges our understanding of distance, location, and anecdote. While traditional GIS is assumed to use geometrically determined systems of distance and location, and 'anecdotal' knowledge is stored within metadata, qualitative GIS codes images, anecdotes, and coordinates in ways which exceed systematization. Again, qualitative GIS reconfigures these staples of GIS, to produce knowledges differently. There is no whole story provided by the qualitative GIS, but only a partial and situated storytelling. Knigge and Cope's research (this volume) is emblematic of this kind of reconceptualization (see also Knigge and Cope, 2006). Their recursive analysis with grounded visualization is about situating knowledges and configuring the GIS to bring these multiple knowledges into collusion. The dissonance created by these juxtapositions, and of these reversals of coordinates, images, and anecdotes, provides for them the necessary elements of 'strong conclusions' (2006: 2021). Their qualitative GIS resists the hegemony of flat cartography, by demonstrating how these cartographies are always already interpolated by databases, images, imaginations, and narrative. Qualitative GIS is not only about placing numbers in context, as Moss (1995) has proposed, but also about allowing these numeracies to mingle with the non-numeric.

The techno-positionality of qualitative GIS engages in knowledge-making practices through the mixing of methods and analysis, both to create different knowledges and to permanently alter the technology, materially and discursively. The qualitativeness of such an endeavor underlines its resistance to prevalent discourses that associate GIS with quantification, logical positivism, and technophilia. However, this agenda is not solely being advanced in the academy, as practitioners are already looking to the next mashup and the next widget to capture the qualities of lives lived in Google Earths. Qualitative GIS, and its techno-positionality, are implicated in global capital; see, for instance, the acquisitions made in online mapping technologies (Francica, 2007). It becomes primarily our responsibility, as academics, to continue to open these movements to interrogation and to consider the shifts in discourse that such technologies and techno-positionalities are enabling. There will be new combinations, and new ways of juxtaposing. As a form of critical GIS research, qualitative GIS should continue to inquire about these emerging ways of creating knowledge, to ask: what is symptomatically not seen in this mode of visuality? Who cannot protest? How is this sort of techno-positionality productive of new sightings and silences?

CONCLUSIONS: A GENEALOGY OF QUALITATIVE GIS?

Qualitative GIS is a kind of borderland, marking a space of in-betweenness among GISciences, high-stakes capitalism, and academic disciplinarity – with always shifting modes of practice and theorization. It seeks to make knowledge differently, by opening our accepted ways of knowing to critique. Like the mixed media piece in Chagoya's *Borderlandia* exhibit with which I opened this chapter, qualitative GIS makes use of multiple registers of knowledge production, in a mix of representative techniques. Chagoya's work is, digitally speaking, mashup. He explores, in juxtaposition, the violent limits and injustices of modernity and modern representation. Chagoya's art works through dissonance and epistemological irreducibles. Similarly, the mixing of qualitative GIS serves to alter the original techniques, to engage in technocritique. Intrigued by the commonalities of these two forms, I have explored here how this technocritique works, and the research positionalities that are constituted through its workings. This is the task of situating qualitative GIS, in the disciplinary trajectories of GIS and society studies, and most specifically critical GIS. It is also the task of imagining qualitative GIS as kin to broader developments in distributed mapping, web GIS, volunteered geographic information (VGI), Google Maps and Google Earth, and emerging geotagging innovations linking fleshly lives to digital ones.

To situate qualitative GIS, I have placed the branches of GIS and society studies in relation to qualitative GIS, to ask how each, as a project, is (dis)similar to qualitative GIS. More specifically, I am interested in the continued project of critical GIS. Therefore, I have attempted to site qualitative GIS within critical GIS, and have proposed that qualitative GIS researchers must continue to problematize their positionality *vis-à-vis* the technology. A genealogy of qualitative GIS must consider how critical GIS, as first articulated by Schuurman (1999) to frame researcher–technology relationships through an 'insider gaze', is now a limiting framework for a qualitative GIS. As a critical GIS, qualitative GIS inherits the anxieties surrounding internal critique, and is bolstered by an insistence on *relevance to* and *legibility through* the technology. Qualitative GIS also shares certain proclivities with STS studies, ethno(carto)graphies, and social-behavioral studies of GIS. As a movement in GIS research and development, qualitative GIS moves beyond this earlier critical GIS in productive ways – in working the tensions of context and contest, images and imaginings, and protocol and protest.

By recognizing this earlier critical GIS as a disciplinary strategy that is timed and placed, we can better articulate how qualitative GIS can address these earlier critiques of GIS while constituting alternative forms of technology. As qualitative GIS researchers, we must recognize the tracings of earlier debates, including insider–outsider ones, that permeate our contemporary projects. Qualitative GIS invokes a different positionality from other modes of GIS and society research, what I have termed techno-positionality. This is the positionality of the conflicted insider, of the performing of research that is simultaneously about and with the technology. This techno-positionality disengages hegemonic practices and representations that permeate the technology. It is a positionality for enacting change through code and through storytelling; it takes discursive materiality seriously and is open to new configurations and problematizations. It is a technology of the borderlands, responsible for the muddying of boundaries and the mixing of methods. These are the messes inhabited by qualitative GIS research – a critical inhabiting to alter our conventions for knowing.

ACKNOWLEDGEMENTS

I thank the editors of this collection for their thoughtful inclusiveness, which speaks to their talents as critical and engaged scholars. Special thanks go to my mother, who enthusiastically suggested a trip to the Des Moines Art Museum, which brightened our bleary holiday weekend. I also thank Bonnie Kaserman, Marianna Pavlovskaya, and the anonymous reviewers for their suggestions. I, of course, take final responsibility for these ideas.

NOTES

1 Thanks to the editors for suggesting 'qualitativeness' here, as it appropriately characterizes the multiple interventions that qualitative GIS enables.
2 Thanks to an anonymous reviewer for this point.
3 See, for example, the handful of paper sessions at the 2007 AAG Meeting on this topic, including 'Google Earth as the "View from Nowhere"', organized by Martin Dodge and Chris Perkins; 'Virtual Globes', organized by Josh Bader and J. Alan Glennon; 'Visualization and Map Communications', chaired by Molly Holmberg; and 'Mapping and the Internet', chaired by Ron McChesney.

REFERENCES

Aitken, S.C. and Craine, J. (2006) 'Guest Editorial: Affective Geovisualizations', *Directions Magazine*, 7 February.

Bell, S. and Reed, M. (2004) 'Adapting to the machine: integrating GIS into qualitative research', *Cartographica*, 39 (1): 55–66.

Benson, K. and Nagar, R. (2006) 'Collaboration as resistance? Reconsidering the processes, products, and possibilities of feminist oral history and ethnography', *Gender, Place and Culture*, 13 (5): 581–92.

Brown, M. and Knopp, L. (2008) 'Queering the map: the productive tensions of colliding epistemologies', *Annals of the Association of American Geographers*, 98 (3): 1–19.

Chrisman, N.R. (2005) 'Full circle: more than just social implications of GIS', *Cartographica*, 40 (4): 23–35.

Crampton, J.W. and Krygier, J. (2005) 'An introduction to critical cartography', *ACME: An International E-Journal for Critical Geographies*, 4 (1): 11–33.

Curry, M.R. (1998) *Digital Places: Living with Geographic Information Technologies*. London: Routledge.

Dennis, S.F. Jr (2006) 'Prospects for qualitative GIS at the intersection of youth development and participatory urban planning', *Environment and Planning A*, 38: 2039–54.

Dragicevic, S. and Balram, S. (2004) 'A web GIS collaborative framework to structure and manage distributed planning processes', *Journal of Geographical Systems*, 6: 133–53.

Elwood, S.A. (2006) 'Beyond cooptation or resistance: urban spatial politics, community organizations, and GIS-based spatial narratives', *Annals of the Association of American Geographers*, 96 (2): 323–41.

Elwood, S.A. and Martin, D.G. (2000) '"Placing" interviews: location and scales of power in qualitative research', *The Professional Geographer*, 52 (4): 649–57.

England, K. (1994) 'Getting personal: reflexivity, positionality and feminist research', *The Professional Geographer*, 46 (1): 80–9.

England, K. (2002) 'Interviewing elites: cautionary tales about researching women managers in Canada's banking industry', in P. Moss (ed.), *Feminist Geography in Practice: Research and Methods*. Oxford: Blackwell. pp. 200–13.

Francica, J. (2007) 'How much is location technology worth?', *Directions Magazine*, 7 October.

Ghose, R. (2007) 'Politics of scale and networks of association in PPGIS', *Environment and Planning A*, 39 (8): 1961–80.

Goodchild, M.F. (2007) 'Call for participants: specialist meeting on volunteered geographic information', http://www.ncgia.ucsb.edu/projects/vgi/, accessed August 2007.

Hannah, M.G. and Strohmayer, U. (2001) 'Anatomy of debate in human geography', *Political Geography*, 20: 381–404.

Haraway, D.J. (1991) *Simians, Cyborgs, and Women: The Reinvention of Nature*. New York: Routledge.

Harris, L.M. and Harrower, M. (2005) 'Critical interventions and lingering concerns: critical cartography/GISci, social theory, and alternative possible futures', *ACME: An International E-Journal for Critical Geographies*, 4 (1): 1–10.

Harvey, F. and Chrisman, N.R. (1998) 'Boundary objects and the social construction of GIS technology', *Environment and Planning A*, 30: 1683–94.

Herbert, S. (2000) 'For ethnography', *Progress in Human Geography*, 24 (4): 550–68.

Jankowski, P. and Nyerges, T.L. (2001) *Geographic Information Systems for Group Decision Making: Towards a Participatory Geographic Information Science*. London: Taylor and Francis.

Johnson, J.T., Louis, R.P. and Pramono, A.H. (2005) 'Facing the future: encouraging critical cartographic literacies in indigenous communities', *ACME: An International E-Journal for Critical Geographies*, 4 (1): 80–98.

Johnston, J. (1999) 'Machinic vision', *Critical Inquiry*, 26 (1): 27–48.

Jung, J.-K. (2007) 'Computer-aided qualitative GIS (CAQ-GIS): a new approach to qualitative GIS', paper presented at the 103rd Annual Meeting of the Association of American Geographers, San Francisco, California.

Katz, C. (1994) 'Playing the field: questions of fieldwork in geography', *The Professional Geographer*, 46 (1): 67–72.

Knigge, L. and Cope, M. (2006) 'Grounded visualization: integrating the analysis of qualitative and quantitative data through grounded theory and visualization', *Environment and Planning A*, 38: 2021–37.

Kwan, M.-P. (2002) 'Feminist visualization: re-envisioning GIS as a method in feminist geographic research', *Annals of the Association of American Geographers*, 92 (4): 645–61.

Kwan, M.-P. (2007) 'Affecting geospatial technologies: toward a feminist politics of emotion', *The Professional Geographer*, 59 (1): 27–34.

Kwan, M.-P. and Knigge, L. (2006) 'Doing qualitative research using GIS: an oxymoronic endeavor?', *Environment and Planning A*, 38: 1999–2002.

Lake, R.W. (1993) 'Planning and applied geography: positivism, ethics, and geographic information systems', *Progress in Human Geography*, 17 (3): 404–13.

Latour, B. (2005) *Reassembling the Social: An Introduction to Actor-Network Theory*. Oxford: Oxford University Press.

Lawson, V. (1995) 'The politics of difference: examining the quantitative/qualitative dualism in post-structuralist feminist research', *The Professional Geographer*, 47 (4): 449–57.

Leszczynski, A. (2007) 'Critique and its discontents: GIS and its critics in postmillennial geographies'. Unpublished MA thesis, Simon Fraser University, Burnaby, British Columbia.

Marshall, J. (2002) 'Borderlands and feminist ethnography', in P. Moss (ed.), *Feminist Geography in Practice: Research and Methods*. Oxford: Blackwell. pp. 174–86.

Matthews, S.A., Detwiler, J.E. and Burton, L.M. (2005) 'Geo-ethnography: coupling geographic information analysis techniques with ethnographic methods in urban research', *Cartographica*, 40 (4): 75–90.

Miller, C.C. (2006) 'A beast in the field: the Google Maps mashup as GIS/2', *Cartographica*, 41 (3): 187–99.

Moss, P. (1995) 'Embeddedness in practice, numbers in context: the politics of knowing and doing', *The Professional Geographer*, 47 (4): 442–9.

Nyerges, T.L., Jankowski, P., Tuthill, D. and Ramsey, K.S. (2006) 'Collaborative water resource decision support: results of a field experiment', *Annals of the American Academy of Political and Social Science*, 96 (4): 699–725.

Pain, R. (2004) 'Social geography: participatory research', *Progress in Human Geography*, 28 (5): 652–63.

Pain, R., MacFarlane, R., Turner, K. and Gill, S. (2006) '"When, where, if, and but": qualifying GIS and the effect of streetlighting on crime and fear', *Environment and Planning A*, 38: 2055–74.

Pavlovskaya, M. (2006) 'Theorizing with GIS: a tool for critical geographies?', *Environment and Planning A*, 38: 2003–20.

Peng, Z.-R. (2001) 'Internet GIS for public participation', *Environment and Planning B: Planning and Design*, 28 (6): 889–905.

Pickles, J. (1997) 'Tool or science? GIS, technoscience, and the theoretical turn', *Annals of the Association of American Geographers*, 87 (2): 363–72.

Pickles, J. (2004) *A History of Spaces: Cartographic Reason, Mapping, and the Geo-coded World*. New York: Routledge.

Poore, B.S. and Chrisman, N.R. (2006) 'Order from noise: towards a social theory of geographic information', *Annals of the Association of American Geographers*, 96 (3): 508–23.

Pratt, G. (1996) 'Trashing and its alternatives', *Environment and Planning D: Society and Space*, 14 (3): 253–6.

Rose, G. (1997) 'Situating knowledges: positionality, reflexivities and other tactics', *Progress in Human Geography*, 21 (3): 305–20.

Schuurman, N. (1999) 'Critical GIS: theorizing an emerging science', *Cartographica*, 36 (4): 1–108.

Schuurman, N. (2000) 'Trouble in the heartland: GIS and its critics in the 1990s', *Progress in Human Geography*, 24 (4): 569–90.

Schuurman, N. (2006) 'Formalization matters: critical GIS and ontology research', *Annals of the Association of American Geographers*, 96 (4): 726–39.

Schuurman, N. and Leszczynski, A. (2006) 'Ontology-based metadata', *Transactions in GIS*, 10 (5): 709–26.

Schuurman, N. and Pratt, G. (2002) 'Care of the subject: feminism and critiques of GIS', *Gender, Place and Culture*, 9 (3): 291–9.

Sheppard, E. (2001) 'Quantitative geography: representations, practices, and possibilities', *Environment and Planning D: Society and Space*, 19: 535–54.

Smith, N. (1992) 'History and philosophy of geography: real wars, theory wars', *Progress in Human Geography*, 16: 257–71.

Staeheli, L.A. and Martin, P.M. (2000) 'Spaces for feminism in geography', *Annals of the American Academy of Political and Social Science*, 571: 135–50.

Williams, C. and Dunn, C.E. (2003) 'GIS in participatory research: assessing the impact of landmines on communities in north-west Cambodia', *Transactions in GIS*, 7 (3): 393–410.

10

CONCLUSION: FOR QUALITATIVE GIS

Meghan Cope and Sarah Elwood

INTRODUCTION

From its inception, GIS has been more than only quantitative, and it is now more open than ever to incorporating qualitative data and intersecting with analyses, epistemologies, and theory building drawn from qualitative research. Similarly, qualitative research increasingly incorporates visual representations of space and place, spatially referenced data, and knowledge production practices drawn from cartography, spatial analysis, and geovisualization. Through this collection, we have sought to illustrate some of these creative collisions and to conceptualize the emerging field of qualitative GIS. But the representational, analytical, and conceptual engagements profiled here are just the beginning. These productive collisions of GIS with qualitative research continue to grow, fueled by the creativity and skills of researchers, by technological innovations, and by the expanding prominence of research paradigms such as participatory action research that prioritize flexibility and accessibility in the processes and products of our inquiry.

PRACTICE AND METHOD

Through the projects described in these chapters, the conceptual edges of qualitative GIS begin to emerge, and a repeated theme is that qualitative GIS is constituted simultaneously as practice and method. Both GIS and qualitative research are far more than mere tools; rather, they are constructed and performed in ways that blend practice and method, technique and epistemology, at every stage of research. Thus, when combined in qualitative GIS, they enable new realms of intersection within the social practices of research, such as the ways researchers and research participants make choices about which tools to use, what constitute 'data', whose voices and perspectives get included or excluded, how displays and other representations are put together, the means of communicating results, and what might be the broader implications and possibilities for new forms of knowledge.

Critical GIS scholars have long identified issues of GIS-as-practice as ripe for critique. For instance, some critics have claimed that GIS has been falsely assumed by some users to be a neutral piece of software independent of social and cultural perspectives, values, power dynamics, masculinist world-views, or service to capitalism and military might. Similarly, qualitative research has often come under fire for being anecdotal, unscientific, or otherwise 'fluffy'. The chapters presented in this volume offer one strong response to these critiques by demonstrating a critically aware engagement with GIS and qualitative research as methods *and* practices that are always already imbued with power. This is one point at which the mixed methods approach that underlies qualitative GIS becomes more than the sum of its parts: because different methods and their associated practices are reflexively blended, the process of knowledge production can tend to become more transparent.

Thus, mixed methods research such as qualitative GIS can enhance the rigor of knowledge production, not only because of its reflexive, critical traditions, but also because a critical mixed methods approach makes it harder for a researcher to be complacent about a single version of 'the truth'. When different data sources and methods of analysis are played off each other and collide in different ways, they result in complementary, contradictory, or subtly nuanced and intersecting *multiple* truths, each of which can be thoroughly explored to foster more robust explanations.

A thoughtful engagement with practices and methods of research also leads us to consider the ways that new forms of knowledge production and new knowledge itself can have influence *back* upon the practices of research, the methodologies, and even the tools themselves. Jung's (this volume) 'computer-aided qualitative GIS', for example, is in one frame a software-level innovation that supports qualitative GIS practice. But simultaneously, his efforts to craft this approach reflect back upon our understanding of the epistemological potential of GIS, as he considers what it is that makes some visual images 'qualitative', and how GIS might be part of inductive qualitative analysis. His approach, then, engages the diverse methodological milieu in which GIS might be situated, and also the inner workings of the technology itself.

Pavlovskaya (this volume) offers another view into the integrative character of qualitative GIS, by renarrating the position of GIS amidst geographers' debates about methodologies, and showing how the forms of data, the computing technologies, and the modes of analysis that are part of GIS can all serve as bases for productive intersections with qualitative methods. This engagement with GIS as tool, as method, *and* as social science critique about knowledge production in research demonstrates how such multi-faceted considerations can make room for qualitative GIS. Knigge and Cope's (this volume) discussion of the potential of grounded visualization for enabling research practice that engages 'scale' as cartographic and social – as process, relationship, and representation – is similarly integrative and multi-vocal. Grounded visualization is an engagement with GIS as method, in this case showing how visualization is a way of knowing that enables insight into these many facets of scale. But at the same time, it also shows how insights generated through grounded visualization may be critical to research practice, suggesting revisions or additions to research activities, or new avenues in analysis and interpretation of ethnographic and geovisual data.

These examples are only a few of innumerable ways in which qualitative GIS emerges as practice and method. The chapters in this collection have illustrated how the methods–praxis intersections that characterize qualitative GIS are born of at least two origins: reflexivity upon the processes, products, and politics of *any* research; and a multi-faceted engagement with GIS as a technology for storing and representing spatial information, as social and political practice, and as a socially constituted approach for knowing and making knowledge. From these origins, as the contributing authors so ably demonstrate, qualitative GIS can emerge as a productively eclectic range of intersections between geographic information systems, spatial knowledge, and spatial methodologies.

KNOWING, IN THE FACE OF AMBIGUITY

As Wilson (this volume) so eloquently identifies, qualitative GIS creates, thrives on, and ultimately speaks from a reordering of *messiness*. By blurring boundaries between technologies and methods, weaving together discourse and materiality, raising questions at every point, and pulling different forms of knowledge into new understandings, qualitative GIS simultaneously revels in the broad diversity of truths *and* turns a critical eye toward those very 'truths', whether they are about technology, method, or how things work in the world. In this sense, there is not now, nor can there ever be, a singular qualitative GIS: it is rooted from the start in multiplicity. This does not mean that 'anything goes' – though experimentation is surely encouraged – but means rather that purposeful, critical engagement with different methods and epistemologies opens the field to new processes of knowledge production.

Qualitative GIS is premised upon the notion that understanding multiple subjectivities and experiences is critically important in research, but this foundation carries the risk of being immobilized by a magnitude of ambiguity and relativism. How can we make knowledge claims in the face of such ambiguity? One way to do so is through intentional and thoughtful engagement with the ballet of knowledge production that occurs with mixed methodologies such as qualitative GIS. In this dance, different but simultaneous layers of routine performance are co-present and interact with each other on the stage. Each layer is important, but some may be hidden, or even silenced, strategically while others hold center stage as the dominant discourse. In the next act the positions may be reversed or reordered, allowing new insight into each layer as they are seen in a different light and in new relative positions. But this is not the scripted ballet of professionals; rather, it is the rich chaos of the street, the agora, that serves as stage, and the events and narratives that are revealed are unpredictable. Feminist theory in particular informs our ability to navigate this dance intentionally and productively, because of its simultaneous attentiveness to the nuances of lived experience and the structures of power, and to their intersections and contradictions. Qualitative GIS does this navigation by actively engaging a broad spectrum of methodological possibilities within geospatial technologies, while also taking seriously the many layerings of meaning, positionality, interpretations, and subjectivities that constitute knowledge. And qualitative GIS further shows its critical feminist roots through a constant reflexive concern for those who (*think* they) are directing the show, and an awareness that their choreography does not *dictate* the production, but rather informs the expression and *interpretation* of the ongoing performance.

QUAL-GIS AS A 'SIMULTANEITY OF STORIES-SO-FAR'

Massey's (2005) effort to rethink space, and especially to question the assumed relations between time and space, brings us some useful ideas for incorporating multiple realities and subjectivities into mixed methods research. Particularly apt are her entwined notions of 'coexisting heterogeneity' and the 'simultaneity of stories-so-far'. She says

> What is needed, I think, is to uproot 'space' from that constellation of concepts in which it has so unquestioningly so often been embedded (stasis; closure; representation) and to settle it among another set of ideas (heterogeneity; relationality; coevalness … liveliness indeed) where it releases a more challenging political landscape. (2005: 13)

Massey's engagement is an attempt to envision not only what *is* (to the extent that this can even be identified) but also what was and what might have been, and what cannot be seen because of the social-cultural blinders that prevent us from seeing it. Her comments on 'the map' are especially relevant here:

> What if space is the sphere not of a discrete multiplicity of inert things, even one which is thoroughly interrelated? What if, instead, it presents us with a heterogeneity of practices and *processes?* Then it will be not an already-interconnected whole but an ongoing product of interconnections and not. Then it will be always unfinished and open … This is space as the sphere of a dynamic simultaneity, constantly disconnected by new arrivals, constantly waiting to be determined (and therefore always undetermined) by the construction of new relations. It is always being made and always therefore, in a sense, unfinished (except that 'finishing' is not on the agenda). (2005: 107, original emphasis)

As we suggested in the introduction to this volume, and as Wilson takes up in his reflection on the positionality of qualitative GIS, the approach of qualitative GIS is one in which there are moments of fixity that are in constant negotiation with that which is flexible. Qualitative GIS involves recognizing and using the power of spatial technologies for analysis, representations, even 'gee-whiz' displays to woo powerful advocates, while also using constant reflexive critique to remain open to alternative explanations, different interpretations, and multiple authorings of knowledge. But taking Massey's envisionings to heart, qualitative GIS could also serve well as a method for revealing and understanding a multitude of possibilities, paths taken and not taken, constantly in flux. Her notion of the 'simultaneity of stories-so-far' suggests looking at a slice of space – understanding that social relations and places are constantly affecting each other and changing, yet that there are also ways to build on previous knowledges that come from familiarity and routines. This is the strength of a methodology that is built on open flexibility to heterogeneity while also employing some means of understanding fixity. Such approaches allow us to reflect on those routines that happen just often enough in time–spaces to generate a pattern that we (quite artificially) pin down on a map.

Thus, a community garden, a park, or a boundary negotiated and contested by neighboring communities is each a relational time–space, 'constructed out of the articulation of trajectories' (Massey, 2005: 179), comprising of many stories coming together at one place and moment, relative to their social and physical surroundings. They are the ephemeral accretions of particular combinations of actions and conditions, and can be manifestations of competing desires or orderings, and colliding presences and absences. For the cash-strapped city, a vacant lot is unrealized capital and an *absence* on the tax rolls; for neighbors, the lot is a lost historical site, currently an eyesore, and may represent the *presence* of a gang that uses it for a hangout, a signal of the 'chaos' of the city. But for the 'hidden hand' of neoliberal political economy, a vacant lot is the result of a rational ordering of property markets in time–space, the lot serving a role in its decrepitude as a disciplinary corrective to spur others to perform the 'work' of transforming it into worthy space, at which time it will be snapped up once again, just as planned. Querying, analyzing, or even imagining these different time–spaces requires methodologies that are themselves open to simultaneity, to heterogeneity, and to acceptance of moments of fixity as only that: fixed moments within a context of flux. Qualitative GIS provides the beginnings of such an approach.

SIGNIFICANCE AND IMPLICATIONS OF QUALITATIVE GIS

At the conclusion of this collection, we return to the question of the significance of qualitative GIS. Why do the representational, analytical, and conceptual engagements with GIS that constitute qualitative GIS matter, and to whom? The approaches illustrated in this volume, and expanding rapidly elsewhere, are important because of their attentiveness to and flexible engagement with knowledge, identity, place, and power in research practices *and* in the complicated messy social worlds we strive to understand. These developments in qualitative GIS have implications for several dimensions of inquiry.

First, a strong thread running through the chapters here is the potential for qualitative GIS to facilitate new forms of knowledge, new practices of knowledge production, and perhaps new ways of knowing, through its capacity as a form of reflexive mixed methods research. In qualitative GIS, cartographic and ethnographic methods blend, weave, and bump up against each other, allowing new ways to understand phenomena such as boundaries, nodes, clusters, flows, patterns, and scale that have long been central to geographers' inquiry. From these new orientations, researchers can produce more fluid, contextualized, and nuanced results, and create rich process-based explanations.

These blendings can also bring together forms of knowledge – particularly from marginalized groups – that have not traditionally been deemed worthy of inclusion in formal mappings. This raises a second dimension of significance for qualitative GIS: identity matters. Radical representations, revelatory cartographies of the oppressed, and participatory engagements with knowledge and knowledge making foster creative explorations of diverse identities and the understandings they produce. From our own work with low-income, racially marginalized urban communities in Buffalo and Chicago, USA, we have observed, participated in, and facilitated these processes. So, when a young girl on Buffalo's West Side mapped the most important sites in her neighborhood as part of the Children's Urban Geographies project (see Jung, this volume; and Cope, 2008), she imbued her own map with her identity as a Puerto Rican, bilingual, Catholic girl with a tight-knit family, recently arrived in the city. Simultaneously, she inscribed her knowledge of that place (a shortcut to a friend's house, the best spot to get candy, the street corner she was told to avoid) into the research project's broader explanations of the neighborhood. Qualitative GIS was not necessarily *required* to accomplish these connections and juxtapositions, but using participatory geoethnography with young people and diverse mappings of 'official' data through a purposeful, mindful mixing of methods helped construct a more nuanced understanding of this neighborhood and opened new readings of identity and place.

The third dimension of significance for qualitative GIS is related to the last point regarding identity: qualitative GIS reflects and builds upon an enduring concern for place. By embracing diverse geographic knowledges and identities and using GIS or other mappings for mixed layers of analysis and representation, researchers find that the question and rich potential of exploring the production of place, place meanings, and contestations over place rapidly blossom. Of course, such explorations have been occurring for a long time, often without methodologies that use spatial technologies. But we suggest that qualitative GIS can foster new ways of looking at place because of its relational, reflexive, and integrative practices. Approaches illustrated in this volume, including grounded visualization, computer-aided qualitative GIS, and affective geovisualization, demonstrate a few of many possibilities for creative and nuanced inquiry into place and space.

Finally, the significance of qualitative GIS also lies in its implications for power. Power relations matter at every stage of research, and the increasing popularity of participatory research practices suggests a substantial movement towards engaging directly with those relations through ethnographic and other qualitative methods. Critical GIS scholars (see Pavlovskaya, Wilson, this volume) skewered some GIS practices for their seeming obliviousness to intrinsic power relations, but more recent developments suggest that many practitioners are thinking more seriously about power issues. Qualitative GIS offers some potential routes for exploring (and challenging) existing power relations that build on the strengths of its constituent traditions, such as critical GIS and feminist epistemology. At one level, using qualitative GIS strategies challenges traditional power relations through its commitment to mixed methods held purposefully in tension with each other, which can reveal the social arenas of oppression, exploitation, discipline, and empowerment through the productive interplay of methods, analysis, and querying. At another level, qualitative GIS challenges power relations precisely because it inherently involves adopting a critical edge, a stance that questions the taken-for-granted assumptions of knowledge and being-in-the-world at every point. And finally, we do not think it is too much of a stretch to suggest that qualitative GIS can – through these practices and principles – emerge as an approach that rewrites the practices of research-as-usual in part because of its capacity for understanding heterogeneous productions of knowledge, identity, place, and power.

QUAL-GIS FUTURES

In this collection, we have sought to illustrate that qualitative GIS is above all a framework that facilitates productive engagements with the myriad multiplicities that comprise geographic information systems. The most salient characteristics of this approach are that very different conceptualizations, representations, or modes of analysis may exist in productive tension, and that researchers might intentionally interplay them to come to new understandings and more robust explanations. For example, Knigge and Cope show

how the inductive iterative analysis practices of grounded visualization – an analytical engagement with qualitative GIS – can engage scale in GIS as both a cartographic representation and a socio-political construction. Elwood, writing on qualitative approaches within grassroots GIS practice, demonstrates how cartographic representations generated in a GIS might be engaged to produce multiple and different understandings of neighborhood, negotiating the meanings or characteristics associated with it as flexible *and* as fixed, or engaging it as material space *and* as imagined space. Aitken and Craine consider the productive potential of rereading GIS through conceptual frameworks that would seem at odds with GIS, showing how a non-representational reading of GIS-based representations facilitates greater insight into the affective and emotive politics that representations may produce.

These collisions of different and potentially contradictory ways of knowing are important for many reasons. They enable us to more fully understand the complicated layerings of GIS practice through which knowledge, power, and politics are created. They provide us examples of ways that we might engage these layerings productively for particular purposes. And they provide a multitude of ways that GIS might be embedded in research in order to tease out the particular *and* the general, the large *and* the small scale, the representational *and* the non-representational, and countless other multiplicities that are part of GIS-based representations, analysis, and epistemologies. But more fundamentally, qualitative GIS takes seriously the feminist notion that carefully and thoughtfully incorporating multiple ways of knowing is some of the most important political work we can do.

The future of qualitative GIS will not, however, be written – or read – by geographers alone. This volume draws from cutting-edge research in such diverse fields as human ecology, landscape architecture, planning, sociology, environmental psychology, and anthropology, and, we hope, will make contributions back to those disciplines, and beyond. Interdisciplinary work and critical theory have increasingly paid attention to space and place in various ways, from using the mapping tools of GIS to employing spatial metaphors such as margin, location, cartographies, and place in critical analysis. The potential of qualitative GIS lies partly in new mixed methods techniques of data production and analysis, but also in conceptual contributions toward envisioning ways to incorporate multiple realities into scholarship. These strengths, we suggest, transcend disciplinary boundaries. Finally, the future of qualitative GIS also lies in the hands of the web-savvy public. The rapid increase in the integration of spatial and social data through online customizable mapping tools, volunteered geographic information (VGI), and creative 'mashups' of diverse narratives, photographs, journeys, sites, and routes suggests a new realm of popular qualitative GIS is on the horizon, one which we heartily embrace.

REFERENCES

Cope, M. (2008) 'Becoming a scholar-advocate: Participatory research with children', *Antipode*, 40 (3): 428–35.

Massey, D. (2005) *For Space*. London: Sage.

INDEX